SpringerBriefs in Microbiology

For further volumes:
http://www.springer.com/series/8911

Luisa Gouveia

Microalgae as a Feedstock for Biofuels

 Springer

Dr. Luisa Gouveia
Laboratório Nacional de Energia e Geologia
Unidade de Bioenergia
Estrada do Paço do Lumiar, Edifício G
1649-038 Lisboa
Portugal
e-mail: luisa.gouveia@lneg.pt

ISSN 2191-5385   e-ISSN 2191-5393

ISBN 978-3-642-17996-9   e-ISBN 978-3-642-17997-6

DOI 10.1007/978-3-642-17997-6

Springer Heidelberg Dordrecht London New York

Cover design: eStudio Calamar, Berlin/Figueres

Printed on acid-free paper

Springer is part of Springer Science+Business Media (www.springer.com)

Contents

Microalgae as a Feedstock for Biofuels

Abstract This review explains the potential use of the so-called "green coal" for biofuel production. A comparison between microalgae and other crops is given, and their advantages are highlighted. The production of biofuels from microalgae biomass is described, such as the use of algae extracts (e.g. biodiesel from oil, bioethanol from starch), processing the whole biomass (e.g. biogas from anaerobic digestion, supercritical fluid, bio-oil by pyrolysis, syngas by gasification, biohydrogen, jet fuel), as well as the direct production (e.g. alcohols, alkanes). Microalgal biomass production systems are also mentioned, including production rates and production/processing costs. Algae cultivation strategy and the main culture parameters are point out as well as biomass harvesting technologies and cell disruption. The CO_2 sequestration is emphasised due to it's undoubted interest in cleaning our earth. Life cycle analysis is also discussed. The algal biorefinery strategy, which can integrate several different conversion technologies to produce biofuel is highlighted for a cost-effective and environmentally sustainable production of biofuels. The author explains some of the challenges that need to be overcome to ensure the viability of biofuel production from microalgae. This includes the author's own research, the use of microorganism fuel cells, genetic modification of microalgae, the use of alternative energies for biomass production, dewatering, drying and processing. The conclusion of the manuscript is the author's view on the potential of microalgae to produce biofuels; the drawbacks and what should be done in terms of research to solve them; which technologies seem to be more viable to produce energy from algae; and which improvements in terms of microalgae, systems, and technologies should take place to enable the algae to fuels concept a reality.

Keywords Bioenergy production · Biofuels · Biorefinery concept · CO_2 sequestration · Environmental sustainability · Green coal · Life cycle analysis of microalgae · Microalgae · Microalgal biomass production systems · Photobioreactors

L. Gouveia, *Microalgae as a Feedstock for Biofuels*, SpringerBriefs in Microbiology,   1
DOI: 10.1007/978-3-642-17997-6_1, © Luisa Gouveia 2011

1 Introduction

Fuels make up a large share of global energy demand ($\sim 66\%$). The development of CO_2-neutral fuels is one of the most urgent challenges facing in our society, to reduce gaseous emissions and their consequential climatic changes, greenhouse and global warming effects. Biofuel production is expected to offer new opportunities to diversify income and fuel supply sources and can help to reduce the adverse effects of the frequent oil supply crisis, as well as developing long-term replacement of fossil fuels, helping non-fossil–fuel-producing countries to reduce energy dependence. This will in turn promote employment in rural areas, reduce greenhouse gas (GHG) emissions, boost the decarburization of transportation fuels, increase the security of energy supply and promote environmental sustainability.

However, to achieve environmental and economic sustainability, production of fuels should require them to be not only renewable, but also capable of sequestering atmospheric CO_2.

2 Microalgae and Biofuels Production

Microalgae are microscopic photosynthetic organisms that are found in both marine and fresh water environments. Their photosynthetic mechanism is similar to land-based plants, due to a simple cellular structure, and the fact that they are submerged in an aqueous environment, where they have efficient access to water, CO_2 and other nutrients, they are generally more efficient in converting solar energy into biomass. The absence of non-photosynthetic supporting structures (roots, stems, etc.) also favours the microalgae in aquaculture (John et al. 2010).

Microalgae appear to represent the only current renewable way to generate biofuels (Chisti 2007; Schenk et al. 2008). Microalgae biofuels are also likely to have a much lower impact on the environment and on the world's food supply than conventional biofuel-producing crops. When Compared with plants biofuel, microalgal biomass has a high caloric value, low viscosity and low density, properties that make microalgae more suitable for biofuel than lignocellulosic materials (Miao et al. 2004), as well as due their inherently high-lipid content, semi-steady-state production, and suitability in a variety of climates (Clarens et al. 2010).

One unique aspect of algae as compared to other advanced feedstocks is the spectrum of species available for amenability for biofuel production. Various species may be selected to optimize the production of different biofuels. Algae offer a diverse spectrum of valuable products and pollution solutions, such as food, nutritional compounds, omega-3 fatty acids, animal feed, energy sources (including jet fuel, aviation gas, biodiesel, gasoline, and bioethanol), organic fertilizers, biodegradable plastics, recombinant proteins, pigments, medicines, pharmaceuticals, and vaccines (Pulz 2004; Pienkos and Darzins 2009).

Microalgae may soon be one of the Earth's most important renewable fuel crops (Campbell 1997). The main advantages of microalgae are (Campbell 1997; Chisti 2007; Huntley and Redalje 2007; Schenk et al. 2008; Li et al. 2008; Rodolfi et al. 2009; Khan et al. 2009):

- a higher photon conversion efficiency (approximately 3–8% against 0.5% for terrestrial plants), which represents higher biomass yields per hectare) and grow at high rates (e.g. 1–3 doublings/day)
- a higher CO_2 sequestration capacity
- it is able to grow in a liquid medium, with better handling, and can utilize salt and waste water streams (saline/brackish water/coastal seawater), thereby reducing freshwater use
- it utilizes nitrogen and phosphorous from a variety of wastewater sources (e.g. agricultural run-off, concentrated animal feed operations and industrial and municipal wastewaters) providing the additional benefit of wastewater bioremediation
- it uses marginal areas unsuitable for agricultural purposes (e.g. desert and seashore lands) and thereby does not compete with arable land for food production
- production is not seasonal and can be harvested batch-wise nearly all-year-round
- cultures can be induced to produce a high concentration of feedstock (oil, starch, biomass)
- algal biomass production systems can be easily adapted to various levels of operational and technological skills
- it can be cultured without the use of fertilizers and pesticides, resulting in less waste and pollution
- the nitrous oxide released can be minimized when microalgae are used for biofuel production (Li et al. 2008)
- they have minimal environmental impact such as deforestation
- the conversion of light to chemical energy can be responsible for a wide range of fuel synthesis: protons and electrons (for biohydrogen), sugars and starch (for bioethanol), oils (for biodiesel) and biomass (for BTL and biomethane) (Fig. 1), via biochemical, thermochemical, chemical and direct combustion processes (Fig. 2)
- they produce value-added co-products or by-products (e.g. proteins, polysaccharides, pigments, biopolymers, animal feed, fertilizers…).

Photosynthesis drives the first step in the conversion of light to chemical energy and is, therefore, ultimately responsible for the production of feedstock required for all biofuels: synthesis of protons and electrons (for Bio-H_2), sugars and starch (for Bio-ethanol), oils (for Biodiesel) and biomass (for BTL products and Bio-methane) (Hankamer et al. 2007; Costa and Morais 2011).

In the international market, the most technically feasible and commercialised alternative renewable fuel sources are biodiesel and bioethanol. These, respectively, can both replace diesel and gasoline in today's cars with not much or no

Fig. 1 The rule of
photosynthesis in biofuel
production

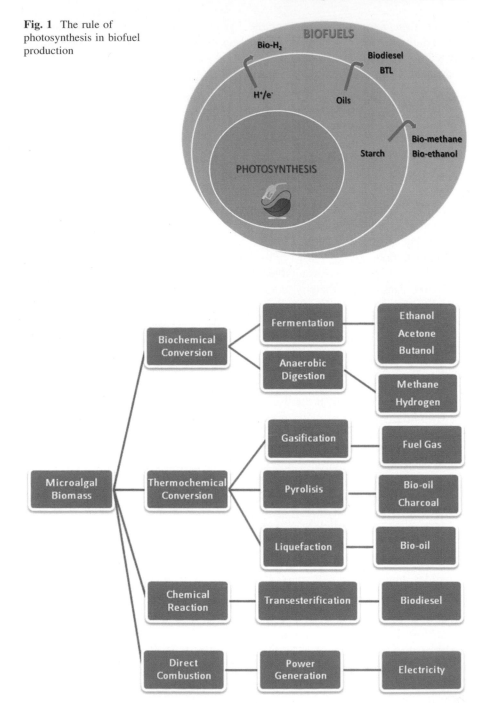

Fig. 2 Energy production by microalgal biomass conversion using biochemical, thermochemical, chemical and direct combustion processes (Wang et al. 2008)

modifications to vehicle engines. They can be produced using existing technologies and can be distributed through the available and existing distribution systems.

When terrestrial biofuels are to replace mineral oil-derived transport fuels, large areas of good agricultural land are needed: about 5×10^8 ha in the case of biofuels from sugar cane or oil palm, and at least $1.8–3.6 \times 10^9$ ha in the case of ethanol from wheat, corn or sugar beet, as produced in industrialized countries (Reijnders 2009).

The overall solar energy conversion efficiency determines net energy yield/ha and this in turn determines land requirements for fossil fuel displacement. In the case of ethanol from sugarcane, the overall solar energy conversion energy efficiency is currently $\sim 0.16\%$ (Kheshgi et al. 2000) and in the case of biodiesel from palm oil $\sim 0.15\%$ (Reijnders and Huijbregts 2009). These percentages are much higher than those for transport biofuels from European wheat and rapeseed (Reijnders 2009).

Both fuels (Biodiesel and Bioethanol) are being produced in increasing amounts as renewable biofuels, but their production in large quantities is not sustainable (Chisti 2007, 2008a, b).

Currently, about 1% (14 million hectares) of the world's available arable land is used for the production of biofuels providing 1% of global transport fuels. Clearly, increasing the share, it will be impractical due to the severe impact on the world's food supply and the large areas of production land required (IEA 2006).

A large number of potential pathways exist for the conversion from algal biomass to fuels. The pathways can be classified into the following three general categories: (1) those that process algal extracts (e.g., lipids, carbohydrates) to yield fuel molecules (e.g. biodiesel, bio-ethanol); (2) those that process whole algal biomass to yield fuel molecules; and (3) those that focus on the direct algal production of recoverable fuel molecules (e.g. ethanol, hydrogen, methane, alkanes) from algae without the need for extraction.

Nevertheless, microalgae that have been pointed as the next feedstock for biofuels due to their very high productivities when compared with the conventional energy crops, many constraints related with harvesting, drying and extraction of oils have delayed the industrial production of microalgal biofuels.

2.1 Algal Extracts

2.1.1 Oils to Biodiesel

Biodiesel is developing into one of the most important near-market biofuels as virtually all industrial vehicles used for farming, transport and trade are diesel based. In the past decade, the biodiesel industry has seen massive growth globally, more than doubling in production every 2 years (Oilworld 2009). Biodiesel represents the highest contribution to the total amount of liquid biofuels produced in the EU with a market share, in 2004, of 79.5%. In Indonesia, Malaysia and

Thailand, biodiesel production rates are currently between 70 and 250% (USDA 2007).

Biodiesel is usually produced from oleaginous crops, such as rapeseed, soybean, sunflower and palm (Al-Widyan and Al-Shyoukh 2002; Sanchez and Vasudevan 2006) through a chemical transesterification process of their oils with short-chain alcohols, mainly methanol. It is a clean burning fuel that can also be used as a substitute or in admixture with diesel, because it is physical and fuel properties are similar and compatible.

However, the increased pressure on arable land currently used for food production could lead to severe food shortages; in particular, for the developing world, where already more than 800 million people suffer from hunger and malnutrition (figure without China) (FAO 2007).

Current biodiesel supplies from oil-producing crops (e.g. sugarcane, sugar beet, maize, sorghum, rapeseed, sunflower, soybean and palm) supplemented with small amounts of animal fat and waste cooking oil, only account for an estimated 0.3% (approximately 12 million tons in 2007) of the global oil consumption (BP statistics 2009), which revealed inefficiency and unsustainability (Chisti 2007; Schenk et al. 2008).

Currently, approximately 8% of plant-based oil production is used as biodiesel, and this has already contributed to an increase in the price of oil crops over the last few years (Oilworld 2009) and cannot even come close to satisfy the existing and future demand for transport fuels. The world's biodiesel industry is presently operating far below capacity due to a lack of feedstock (Schenk et al. 2008).

The second-generation biofuels are derived from non-food feedstock; namely, microalgal and other microbial sources, lignocellulosic biomass, rice straw and bioethers. They are increasingly predicted by international experts and policy-makers to play a crucial role in a clean environmentally sustainable future and a better option for addressing the food and energy security (Schenk et al. 2008; Patil et al. 2008).

The required rapid growth of carbon-neutral renewable alternatives makes microalgae one of the main future sources of biofuels, which could be the only one that could meet the global demand for transport fuels (Chisti 2007; Hu et al. 2008a, b; Schenk et al. 2008).

Several authors have stated that the oil yields from algae are substantially higher than those from oleaginous plants, although the microalgae oil yield is strain dependent. Tickell (2000) presented a possibility to produce around 30 tons biodiesel/ha year, using diatomaceous, considering an average production yield of 50 g/m^2 day containing 50% (w/w) of oil, while with a rapeseed culture the average biodiesel production is one order of magnitude lower (3 tons/ha year). Both yield productivity and oil content values, are quite optimist and not reach so far.

There is a significant variation in the overall biomass productivity and resulting oil yield, land use and biodiesel productivity, however, with a clear advantage for microalgae (Table 1) (Mata et al. 2010).

Table 1 Comparison of microalgae with other biodiesel feedstocks (Chisti 2007; Mata et al. 2010)

Plant source	Seed oil content (%/wt biomass)	Oil Yield (L/ha year)	Land use (m² year/kg biodiesel)	Biodiesel productivity (kg/ha year)
Corn/maize (Zea mays L.)	44	172	66	152
Hemp (Cannabis sativa L.)	33	363	31	321
Soybean (Glycine max L.)	18	636	18	562
Jatropha (Jatropha curcas L.)	28	741	15	656
Camelina (Camelina sativa L.)	42	915	12	809
Canola/rapessed (Brassica napus L.)	41	974	12	862
Sunflower (Helianthus annus L.)	40	1,070	11	946
Castor (Ricinus communis)	48	1,307	9	1,156
Palm (Elaeis guineensis)	36	5,366	2	4,747
Microalgae (low oil content)	30	58,700	0.2	51,927
Microalgae (medium oil content)	50	97,800	0.1	86,515
Microalgae (high oil content)	70	126,900	0.1	121,104

A number of studies have attempted to calculate the cost of algal oil production from large-scale farms. Despite the algae field still being in its infancy and much research still having to be done to reduce costs and improve efficiency, microalgal biodiesel production systems may already be economically viable, even using the existing low-tech approaches (Schenk et al. 2008).

Microalgae strains, such as *Chlorella vulgaris*, *Spirulina maxima*, *Chlamydomonas reinhardtii*, *Nannochloropsis* sp., *Neochloris oleoabundans*, *Scenedesmus obliquus*, *Nitzchia* sp., *Schizochytrium* sp., *Chlorella prototothecoides*, *Dunaliella tertiolecta* among others, were screened by many authors to choose the best lipid producer(s) in terms of quantity (combination of biomass productivity and lipid content) and quality (fatty acid composition) as an oil source for biodiesel production (Xu et al. 2006; Miao and Wu 2006; Chisti 2007; Rodolfi et al. 2009; Lopes da Silva et al. 2009; Gouveia and Oliveira 2009; Gouveia et al. 2009; Morowvat et al. 2010).

Typical microalgal biomass fatty acid composition is mainly composed of a mixture of unsaturated fatty acids, such as palmitoleic (16:1), oleic (18:1), linoleic (18:2) and linolenic (18:3) acid. Saturated fatty acids, such as palmitic (16:0) and stearic (18:0) are also present to a small extent (Meng et al. 2009; Gouveia and Oliveira 2009).

The synthesis and accumulation of a large amount of triacylglycerols (TAG) accompanied by considerable alterations in lipid and fatty acid composition, occurs under stress imposed by chemical and physical environmental stimuli, either acting individually or in combination.

The lipid content increases considerably (doubles or triples) when the cells are subjected to unfavourable culture conditions, such as photo-oxidative stress and/or nutrient starvation (Hu et al. 2008a, b; Gouveia et al. 2009). Fatty acid composition can also vary both quantitatively and qualitatively with their physiological status and culture conditions (Hu et al. 2008a, b).

The major chemical stimuli are nutrient (nitrogen, phosphorous, sulphur, silicon for diatoms) starvation, salinity and pH.

The major physical stimuli are temperature and light intensity; the former, can modify the fatty acid composition by increasing the unsaturation with the decreasing temperature and vice versa (Lynch and Thompson 1982; Raison 1986). Low light intensity induces the formation of polar lipids, whereas high light intensity decreases total polar lipid content, with a concomitant increase in the amount of neutral storage lipids, mainly TAGs (Brown et al. 1996a, b; Khotimchenko and Yakovleva 2005).

Low light favours the formation of PUFAs, whereas high light favours the synthesis of more saturated and mono-unsaturated fatty acids that mainly make up neutral lipids (Sukenik et al. 1993).

In addition to chemical and physical factors, growth phase and/or aging or senescence of the culture also affects the TAG content and fatty acid composition (Hu et al. 2008a, b). Lipid content and fatty acid composition are also subject to variation during the growth cycle, usually with an increase in TAGs in the stationary phase. The aging culture also increases the lipid content of the cells, with a notable increase in the saturated and mono-unsaturated fatty acids, and a decrease in PUFAs (Liang et al. 2006).

Other than the chemical and physical stress factors that may change the microalgal biomass lipid content and composition, the extraction method can also significantly affect the lipid yield. Microalgal lipid extraction usually follows two steps: cell disruption (which greatly depends on cell shape, size and wall structure) and solvent extraction (which depends on lipid composition). The lipid extraction method works differently depending on the alga species (Shen et al. 2009).

Some microalgae can provide standard oil similar with other vegetable crops (Table 2) which can meet biodiesel specifications (Guerrero 2009; Amin 2009), being the physical and fuel properties of biodiesel from microalgae oil, in general, comparable to those of Diesel fuel (Table 3). The biodiesel from microalgae oil

Table 2 Comparison of microalgae oil parameters with other crops (Guerrero 2009)

Oil	Viscosity (cP at 40°C)	Combustion heat (kJ/g)
Palma	38	38.30
Canola	33	38.52
Corn	31	
Soya	26	38.37
Microalgae	36.6	38.72

Table 3 Comparison of properties of microalgae biodiesel, diesel fuel and ASTM standard (Amin 2009)

Properties	Biodiesel microalgae oil	Diesel fuel	ASTM biodiesel standard
Density (kg l^{-1})	0.864	0.838	0.86–0.90
Viscosity (mm^2 s^{-1}, cSt at 40°C)	5.2	1.9–4.1	3.5–5.0
Flash point (°C)	115	75	Min 100
Solidifying point (°C)	−12	−50 to 10	–
Cold filter plugging point (°C)	−11	−3.0 (max −6.7)	Summer max 0 Winter max <−15
Acid value (mg KOH g^{-1})	0.374	Max 0.5	Max 0.5
Heating value (MJ kg^{-1})	41	40–45	–
H/C ratio	1.81	1.81	–

showed a much lower cold filter plugging point of −11°C in comparison with that of Diesel fuel, as shown in Table 3.

2.1.2 Starch to Bioethanol

Bioethanol is already well established as a fuel most notably in Brazil and US (Goldemberg 2007). It is usually obtained by alcoholic fermentation of starch (cereal grains, such as corn, wheat and sweet sorghum), sugar (sugar cane and sugar beet) and lignocellulosic feedstocks (Antolin et al. 2002). The extracted starch is usually mixed with water and heated briefly, before adding the enzymes, and can be hydrolyzed to produce the monomeric sugar glucose, which will be readily metabolized to ethanol by the yeast *Saccharomyces cerevisiae* or *Zymomonas mobilis*.

Saccharomyces cerevisiae is the universal organism for fuel ethanol production from glucose. Nevertheless, *Z. mobilis* is considered as the most effective organism for production of ethanol, although it is not currently used commercially (Drapcho et al. 2008). The ethanol is then purified from the mixture by distillation and dehydration.

Starch processing is a mature industry, and commercial enzymes required for starch hydrolysis are currently an available low cost technology.

Microalgal bioethanol can be produced through two distinct processes: via dark fermentation or yeast fermentation.

The dark fermentation of microalgae consists of the anaerobic production of bioethanol by the microalgae itself through the consumption of intracellular starch. The yeast fermentation process is well established industrially and to achieve higher yields, it is necessary to screen strains with high starch and other sugars contents and induce accumulation of intracellular starch.

Despite dark fermentation being a low-energy intensive process, the yields obtained were 1% (w/w) for the strain *Chlamydomonas reinhardtii* (Hirano et al. 1997) and 2.07% (w/w) for *Chlorococum littorale* (Ueno et al. 1998), which does not make this process appealing to the industry. The microalgae *Chlamydomonas perigranulata* has also been reported to produce intracellular bioethanol (Hon-Nami and Kunito 1998; Hon-Nami 2006).

Some microalgae have a high starch content (Table 4), and, therefore, a high potential for bioethanol production which has been mentioned by many authors (e.g. Schenk et al. 2008; Hankamer et al. 2007) however only some research (Huntley and Redalje 2007; Rosenberg et al. 2008; Subhadra and Edwards 2010) has been done on this subject (Douskova et al. 2008).

It has been estimated that approximately 46,760 to 140,290 L of ethanol/ ha year can be produced from microalgae (Cheryl 2010). This yield is several orders of magnitude higher than yields obtained for other feedstocks (Table 5).

Matsumoto et al. (2003) have screened several strains of marine microalgae with high carbohydrate content, and identified a total of 76 strains with a carbo-hydrate content ranging from 40 to 53%.

Hirano et al. (1997) conducted an experiment with *C. vulgaris* microalga (37% starch content) through fermentation and yielded a 65% ethanol-conversion rate, when compared with the theoretical conversion rate from starch. Ueda et al. (1996) found that microalgae, such as *Chlorella*, *Dunaliella*, *Chlamydomonas*, *Scene-desmus*, *Spirulina* contain large amounts (>50%) of starch and glycogen which are useful as raw materials for ethanol production. Microalgae can assimilate cellulose that can be fermented to bioethanol (Chen et al. 2009).

The microalgae *Chlorococum* sp. has also been studied as a feedstock for ethanol production (Harun et al. 2010b). In this study, it was demonstrated that the cell wall disruption improves the yield of the process and that lower biomass concentrations produce greater ethanol concentrations. The maximum productivity

Table 4 Amount of carbohydrates from various species of microalgae on a dry matter basis (%) (Adapted from Becker 1994; Harun et al. 2010a)	Algae strains	Carbohydrates (%/wt biomass)
	Scenedesmus obliquus	10–17
	Scenedesmus quadricauda	–
	Scenedesmus dimorphus	21–52
	Chlamydomonas reinhardtii	17
	Chlorella vulgaris	12–17
	Chlorella pyrenoidosa	26
	Spirogyra sp.	33–64
	Dunaliella bioculata	4
	Dunaliella salina	32
	Euglena gracilis	14–18
	Prymnesium parvum	25–33
	Tetraselmis maculate	15
	Porphyridium cruentum	40–57
	Spirulina platensis	8–14
	Spirulina maxima	13–16
	Synechoccus sp.	15
	Anabaena cylindrical	25–30

Table 5 Ethanol yield from different sources (adapted from Mussatto et al. 2010)

Source	Ethanol yield (L/ha)	References
Corn stover	1,050–1,400	Tabak (2009)
Wheat	2,590	Cheryl (2010)
Cassava	3,310	Cheryl (2010)
Sweet sorghum	3,050–4,070	Lueschen et al. (1991); Hills et al. (1983)
Corn	3,460–4,020	Tabak (2009)
Sugar beet	5,010–6,680	Hills et al. (1983); Brown (2006)
Sugarcane	6,190–7,500	Brown (2006)
Switch grass	10,760	Brown (2006)
Microalgae	46,760–140,290	Cheryl (2010)

was 38% (w/w). The same authors found out that it is essential to incorporate a pre-treatment stage to release and convert the complex carbohydrates entrapped in the cell wall, into simple sugars, before the fermentation process. The highest bioethanol concentration obtained by these authors was 7.20 g/L, achieved when the pre-treatment step was performed with 15 g/L of microalgae at 140°C using 1% (v/v) of sulfuric acid for 30 min. In terms of ethanol yield, a maximum of ~ 52 wt% (g ethanol/g microalgae) was obtained using 10 g/L of microalgae and 3% (v/v) of sulfuric acid under 160°C for 15 min.

Temperature was found to be the most important parameter that influences bioethanol production from microalgae, followed by the acid concentration and the amount of microalgae.

According to Harun and Danquah (2011), an acid pre-treatment is the best pre-treatment before fermentation, as compared to other pre-treatment methods, namely in terms of cost-effectiveness and low energy consumption.

It has been demonstrated that supplementing the medium with iron can increase threefold the carbohydrate content (He et al. 2010). Douskova et al. (2008) have shown that for the microalgae *C. vulgaris* in phosphorus, nitrogen or sulphur limiting conditions, the starch content of the cells increased 83, 50 and 33%, respectively.

The production of bioethanol from the fermentation of microalgal biomass presents itself some advantages because it can use leftover microalgae from other processes (e.g. oil extraction) or intact biomass; it occurs in an aqueous medium; therefore, there is no need to spend energy drying the biomass and the biomass necessary can be concentrated by simply settling. The chemical cell disruption techniques can simultaneously breakdown complex sugars necessary for yeast fermentation and the yeast fermentation technology is well established industrially.

2.2 Processing of Whole Algae

In addition to the direct production of biofuels from algae, it is also possible to process whole algae into fuels instead of first extracting oils and/or starch and

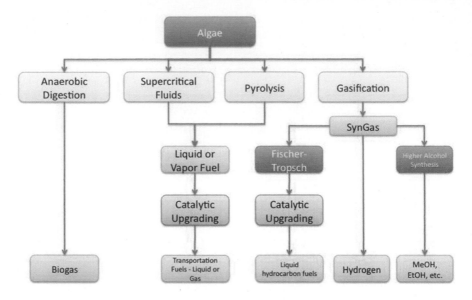

Fig. 3 Schematic of the potential conversion routes for whole algae into biofuels

post-processing. These methods benefit from reduced costs associated with the extraction process, and the added benefit of being amenable to process a diverse consortium of algae, although at least some level of dewatering is still required. However, the global processes should be economically evaluated, because may be used only the residual biomass after oil, starch and high-value products extraction should economically maximize the entire process (Harun et al. 2010a).

There are four major conversion technologies that are capable of processing whole algae: anaerobic digestion, supercritical processing, pyrolysis and gasification (Fig. 3).

2.2.1 Biomethane (Biogas) by Anaerobic Digestion

Organic material such as crop biomass or liquid manure can be used to produce biogas via anaerobic digestion and fermentation. Mixtures of bacteria are used to hydrolyze and break down the organic biopolymers (i.e. carbohydrates, lipids and proteins) into monomers, which are then converted into a methane-rich gas via fermentation (typically 50–75% CH_4). Carbon dioxide is the second main component found in biogas (approximately 25–50%) and, like other interfering impurities, has to be removed before the methane is used for electricity generation.

Microalgae biomass is a source of a vast array of components that can be anaerobically digested to produce biogas. The use of this conversion technology eliminates several of the key obstacles that are responsible for the current high costs associated with algal biofuels, including drying, extraction, and fuel conversion, and as such may be a cost-effective methodology. Several studies have been carried out that demonstrate the potential of this approach. According to Sialve et al. (2009), the methane content of the biogas from microalgae is 7 to 13% higher when compared with the biogas from maize.

A recent study indicated that biogas production levels of 180.4 mL/g day of biogas can be realized using a two-stage anaerobic digestion process with different strains of algae, with a methane concentration of 65% (Vergara-Fernandez et al. 2008). As it is scientifically well known, microalgae biomass composition are directly related with growth conditions; most microalgae have the capacity, under certain conditions, to accumulate important quantities of carbon in the form of starch or lipids (Qiang 2004), being the nitrogen deficiency, a well-known condition to stimulate this accumulation. Calorific value are directly correlated with the lipid content, and under nitrogen starvation results in a significant increase in the caloric value of the biomass with a decrease in the protein content and a reduction in the growth rate (Illman et al. 2000). Based on the data of these authors, Sialve et al. (2009) evaluated the energetic content (normal and N-starvation growth) of the microalgae *C. vulgaris*, *C. emersonii* and *C. protothecoides*, in two scenarios, such as the anaerobic digestion of the whole biomass and of the algal residues after lipids extraction. From the latter process, biodiesel and methane could be obtained with a higher energetic value. However, the energetic cost of biomass harvesting and lipid recovery is probably higher than the recovery energy, especially because most of the techniques involve biomass drying (Carlsson et al. 2007). When the cell lipid content does not exceed 40%, anaerobic digestion of the whole biomass appears to be the optimal strategy on an energy balance basis, for the energetic recovery of cell biomass, as concluded by the authors (Sialve et al. 2009).

Another study performed by Mussgnug et al. (2010), some microalgae were screening, namely *Chlamydomonasreinhardtii*, *Chlorellakessleri*, *Euglena gracilis*, *Spirulina* (*Arthrospira*) *platensis*, *S. obliquus* and *Dunaliella salina*, and it was demonstrated that the quantity of biogas potential is strongly dependent on the species and on the pre-treatment. *C. reinhardtii* revealed being the more efficient with a production of 587 mL (±8.8 SE) biogas/g volatile solids.

For biogas production, the microalgae species should have a high degree of degradation and low amount of indigestible residues (Mussgnug et al. 2010). The substrates should be concentrated but drying process should be avoided, as it result in a general decrease in the biogas production potential in around 20%. This result represents a good one as it saves energy and time. However, to avoid transportation of the wet biomass, the algal production facility and the biogas fermentation plant should be as close as possible (Mussgnug et al. 2010).

Anaerobic digestion well explored in the past, will probably re-emerge in the coming years either as a mandatory step to support large-scale microalgal cultures or as a standalone bioenergy-producing process (Sialve et al. 2009). This technology could be very effective for situations such as integrated wastewater treatment, where algae are grown under uncontrolled conditions using strains that are not optimized for lipid production.

2.2.2 Supercritical Fluid

Supercritical processing is a recent addition to the portfolio of techniques capable of simultaneously extracting and converting oils into biofuels (Demirbas 2007). Supercritical fluid extraction of algal oil is far more efficient than traditional solvent separation methods, and this technique has been demonstrated to be extremely powerful in the extraction of other components within algae (Nobre et al. 2006; Gouveia et al. 2007; Mendes 2008). This supercritical transesterification approach can also be applied for algal oil extracts. Supercritical fluids are selective, thus providing high purity and product concentrations. In addition, there are no organic solvent residues in the extract or spent biomass (Demirbas 2009a, b), although these method are restricted due to process economics concerns (Ehimen et al. 2010). Extraction is efficient at modest operating temperatures, for example, at <50°C, thus ensuring maximum product stability and quality. In addition, supercritical fluids can be used on whole algae without dewatering, thereby increasing the efficiency of the process.

Further study need to be done to avoid saponification when combined extraction and transesterification of algae are performed; also to evaluate the yield, cost, and efficiency of processing the whole algae.

The Mcgyan process, a continuous transesterification, combines alcohol and lipid into a fixed-bed reactor filled with a sulphated metal oxide catalyst at elevated temperature and pressure, with the advantage of reuse catalyst, less reaction time, or water or other dangerous chemicals (Um and Kim 2009; Krohn et al. 2011)

2.2.3 Bio-Oil by Pyrolysis

Pyrolysis is the thermal decomposition of materials in the absence of oxygen or when significantly less oxygen is present than required for complete combustion (Balat 2009), and three phase products are produced: vapour, liquid and solid phases. The liquid phase is a complex mixture, called bio-oil, of which its composition varies significantly depending on the feedstock and processing conditions.

Slow pyrolysis of biomass is associated with high charcoal content (350–700°C), but the fast pyrolysis, where ground fine particles of feedstock are quickly heated to between 350° and 500°C for less than 2 s, is associated with liquid fuels, and/or gas at higher temperature (Encinar et al. 1998). Fast pyrolysis process for conversion of algae biomass is used to produce 60–75% of liquid bio-oil, 15–25% of solid char, and 10–20% of non-condensable gases, depending on the feedstock used (Mohan et al. 2006).

The bio-oil has been demonstrated to be suitable for power generation via external combustion (e.g. steam, organic Rankine and Stirling cycles) and internal combustion (e.g. diesel and gas turbine engines) or by cofiring with fossil diesel or natural gas (Bridgwater et al. 1999; Czernik and Bridgwater 2004; Li et al. 2008; Wang et al. 2008).

Extensive studies have been carried out on biomass conversion; several technologies, such as an entrained flow reactor, circulating fluid bed gasifier (Prins et al. 2006; Zwart et al. 2006; Meier and Faix 1999), vacuum pyrolysis (Pakdel and Roy 1991) and vortex reactor (Meier and Faix 1999) have been demonstrated to be effective. Pyrolysis has one major advantage over other conversion methods, in that it is extremely fast, with reaction times of the order of seconds to minutes.

The overall energy of biomass ratio of a well-controlled pyrolytic process could be as high as 95.5%.

Many companies transform carbon-based feedstocks, either wood "biomass", petroleum hydrocarbons or municipal solid wastes, into more valuable chemical and fuel products. Nevertheless, they have several undesirable features, such as high oxygen content, low heat content, high viscosity at low temperature, and chemical instability (Czernik and Bridgwater 2004; Chiaramonti et al. 2007), that hamper their use as quality transportation fuels. To overcome this limitation, studies have been taken to upgrade the bio-oils to high quality fuels; namely, by gasification and subsequent Fischer–Tropsch synthesis (Lv et al. 2007; Iliopoulou et al. 2007), hydrogen production by steam-reforming bio-oil (Wang et al. 2007).

Recently, a few investigations have been carried out regarding the suitability of microalgal biomass for bio-oil production (Wu and Miao 2003; Miao and Wu 2004; Miao et al. 2004; Dermibas 2006; Grierson et al. 2008).

Miao et al. (2004) studied fast pyrolysis of *Chlorella protothecoides* and *Microcystis aeruginosa* and they reported bio-oil yields of 18% [higher heating value (HHV) of 30 MJ/kg] and 24% (HHV of 29 MJ/kg), respectively.

Dermibas (2006) has shown that, in general, microalgae bio-oils are of much higher quality than bio-oil from wood. Grierson et al. (2008) used a dried and finely ground *Tetraselmis* and *Chlorella* species algae biomass using a slow pyrolysis method and found 43% (v/v) of heavy bio-oil could be produced.

At "flash" pyrolysis, algae have major advantage over other biomass sources due to its inherent small size and the absence of fibber tissue.

2.2.4 Fuel Gas or Syngas by Gasification

This combustible gas mixture is the product of the gasification of biomass, at high temperature (~ 800 to $900°C$) by the partial oxidation of biomass with air, oxygen and/or steam (Wang et al. 2008).

The low-calorific value gas produced (~ 4–6 MJ N/m^3) can be burnt directly for heating or electricity generation, or used as a fuel for engines and gas turbines.

Gasification of the algal biomass may provide an extremely flexible way to produce different liquid fuels, primarily through Fischer–Tropsch synthesis (FTS) or mixed alcohol synthesis of the resulting syngas. FTS is also a relatively mature technology, where the syngas components (CO, CO_2, H_2O, H_2, and impurities) are cleaned and upgraded to usable liquid fuels through a water–gas shift and CO hydrogenation (Okabe et al. 2009).

Hirano et al. (1998) studied gasification of *Spirulina* at temperatures ranging from 850 to 1,000°C for methanol production. They estimated that algae biomass gasification at 1,000°C produced the highest theoretical yield of 0.64 g methanol/g of algae biomass.

Dote et al. (1994) found that thermochemical liquefaction of microalgae species such as *Botryococcus braunii*, *D. tertiolecta* and *Spirulina platensis* yielded 64, 42 and 30–48% (dry wt. basis) of oil and fuel properties of biocrude oil (30–45.9 MJ/kg), which was close to that of petroleum-based heavy oil (42 MJ/kg) (Jena and Das 2009).

The bio-oil yields for the microalgae are 5–25 wt.% lower than the yields of bio-crude, and depending on the biochemical composition. The yields of bio-crude follow the trend lipids > proteins > carbohydrates (Biller and Ross 2011).

Conversion of bio-syngas has several advantages to other methods. First and foremost, it is possible to create a wide variety of fuels with acceptable and known properties. In addition, bio-syngas is a versatile feedstock, and it can be used to produce a number of products, making the process more flexible. Another advantage is the possibility to integrate an algal feedstock into an existing thermochemical infrastructure and wet biomass could be processed (Clark and Deswarte 2008).

It may be possible to feed algae into a coal gasification plant to reduce the capital investment required and improve the process efficiency through economy of scale. In addition, because FTS is an exothermic process, it should be possible to use some of the heat for drying the algae during a harvesting/dewatering process (for other applications) with a regenerative heat exchanger.

Another interesting approach would be the study of the feasibility using the oxygen generated by algae for the use in the gasifier to reduce or eliminate the need for a tar reformer.

2.2.5 Bio-Hydrogen

Hydrogen is an energy carrier with a great potential in the transport sector, for domestic and industrial applications, where it is being explored for liquefaction of coal and upgrading of heavy oils in order to use in combustion engines, and fuel–cell electric vehicles (Balat 2005), and generates no air pollutants. Hence, in both the near and long term, hydrogen demand is expected to increase significantly (Balat 2009).

Hydrogen can be produced in a number of ways (Madamwar et al. 2000). However, currently, the developing H_2 economy is almost entirely dependent on the use of carbon-based non-renewable resources, such as steam reforming of natural gas ($\sim 48\%$), petroleum refining ($\sim 30\%$), coal gasification ($\sim 18\%$) and nuclear powered water electrolysis ($\sim 4\%$) (Gregoire-Padro 2005); from water through thermal and thermochemical processes, such as electrolysis and photolysis; and through biological production, such as steam reforming of bio-oils (Wang et al. 2007), dark and photo fermentation of organic materials and photolysis of water catalyzed by special microalga and cyanobacteria species (Kapdan and Kargi 2006).

Algal biomass (whole or after oil and/or starch removal) can be converted in bio-H_2 by dark-fermentation that is one of the major bio-processes using anaerobic organisms for bio-H_2 production. *Enterobacter* and *Clostridium* strains bacteria are well known as good producers of bio-H_2 that are capable of utilizing various types of carbon sources (Angenent et al. 2004; Das 2009; Cantrell et al. 2008).

On the other way, cyanobacteria and green algae are the only organisms currently known to be capable of both oxygenic photosynthesis and bio-H_2 production.

In cyanobacteria, hydrogen is produced by a light-dependent reaction catalyzed by nitrogenase or in dark-anaerobic conditions by a hydrogenase (Rao and Hall 1996; Hansel and Lindblad 1998), while in green algae, hydrogen is produced photosynthetically by the ability to harness the solar energy resource, to drive H_2 production, from H_2O (Melis et al. 2000; Ghirardi et al. 2000; Melis and Happe 2001; Ran et al. 2006; Yang et al. 2010).

To date, H_2 production has been observed in only 30 genera of green algae (Boichenko and Hoffmann 1994) highlighting the potential to find new H_2-producing eukaryotic phototrophs with higher H_2 producing capacities.

For photobiological H_2 production, cyanobacteria, formerly called "blue green algae" and "nitrogen-fixing" bacteria, are among the ideal candidates, because they have the simplest nutritional requirements. They can grow using air, water and mineral salts, with light as their only source of energy (Tamagnini et al. 2007; Lindblad et al. 2002).

In fact, cyanobacteria (mainly its mutants) are considered the highest biological producer at low cost, since they require only air (N_2 or CO_2), water and mineral salts, using light as the only energy source.

In what concern molecular and physiology H_2 production by cyanobacteria two enzymes are involved: the nitrogenase(s) and the bi-directional hydrogenase. In N_2-fixing strains, the net H_2 production is the result of H_2 evolution by nitrogenase and H_2 consumption mainly catalysed by an uptake hydrogenase. Consequently, the production of mutants deficient in H_2 uptake activity is necessary. Moreover, the nitrogenase has a high ATP requirement and this lowers considerably its potential solar energy conversion efficiency. On the other hand, the bi-directional hydrogenase requires much less metabolic energy, but it is extremely sensitive to oxygen (Das and Veziroglu 2001; Schütz et al. 2004).

Masukara et al. (2001) demonstrated that uptake-deficient mutants of *Anabaena* strains produce considerably more H_2 as compared to the wild types.

The maximum light-driven H_2 production rates for cyanobacteria have been reported to be 2.6 mmol H_2/g h cultured in an Allen and Arnon culture medium, with nitrate molybdenium replaced by vanadium at 30°C, at 6,500 lux of irradiance and 73 Ar, 25 N_2 and 2 CO_2 (% vol.) (Stage I). At a stage II, gas atmosphere was changed to 93 Ar, 5 N_2 and 2 CO_2 (% vol.) (Sveshnikov et al. 1997). These rates compare well with the hydrogen production from the most active green algae at 0.7–1 mmol H_2/g sustainable for hours to days in wild type strains of *Chlamydomonas reinhardtii, C. vulgaris* and *S. obliquus* (Boichenko et al. 2004).

Coupling H_2 Production to Desalination

The use of marine algae has the theoretical potential to couple bio-H_2 production to desalination. Marine and halophilic algae can extract hydrogen (as protons and electrons) and oxygen from sea water and upon combustion of hydrogen and oxygen, fresh water is produced, although at relatively low production rates. This approach requires stationary fuel cells that use hydrogen and oxygen to feed electricity into the national grid (Hankamer et al. 2007), so energy generation can be coupled with desalination. Although the fresh water yield is not large, it does provide a net fresh water gain, whereas conventional crops do not. The yield of water is directly related to the yield of H_2. At 1% light to H_2 efficiency (the approximate current status at outside light levels using light dilution reactors), our feasibility study indicates that upon successful development of the process, a 1 million litre photo-bioreactor facility could produce up to 610,000 L of fresh water per year (Schenk et al. 2008).

Coupling H_2 Production to Carbon Sequestration

Hydrogen is unique among biofuels in that it is carbon free. Consequently, during H_2 production, the vast majority of the CO_2 sequestered during the aerobic phase remains in the residual biomass at the end of the process. By converting this biomass to Agri-char, H_2 fuel production can, therefore, be coupled to atmospheric and industrial CO_2 sequestration. Because carbon trading schemes come on stream internationally, this would add considerable value to the overall bio-H_2 process (Schenk et al. 2008).

Important CO_2 sequestration, advances in fuel cell technology and the fact that the combustion of H_2 produces only H_2O further increases the attractiveness of bio-H_2 production. However, the future of biological hydrogen production depends not only on the research advances, i.e. improvement in efficiency through genetically engineered algae and/or the development of advanced photobioreactors, but also on economic considerations, social acceptance, and the development of a robust hydrogen infrastructure throughout the world.

2.2.6 Jet Fuel

Commercial application of algae-derived jet fuel was further buttressed when, on January 8th, 2009, *Continental Airlines* ran the first test for the first flight of an algae-fueled jet. The test was done using a twin-engine commercial jet, consuming a 50/50 blend of biofuel and normal aircraft fuel. A series of tests executed at

38,000 ft (11.6 km), including a mid-flight engine shutdown, showed that no modification to the engine was required. The fuel was praised for having a low-flash point and sufficiently low-freezing point, issues that have been problematic for other biofuels (BBC 2009).

2.3 Direct Production

The direct production of biofuel from algal biomass has certain advantages in terms of process cost because it eliminates several process steps (e.g., extraction) and their associated costs in the overall fuel production process. There are several biofuels that can be produced directly from algae, including alcohols, alkanes, and hydrogen.

2.3.1 Alcohols

Algae, such as *C. vulgaris* and *C. perigranulata*, are capable of producing ethanol and other alcohols through heterotrophic fermentation of starch (Hon-Nami 2006; Hirayama et al. 1998). This can be accomplished through the production and storage of starch through photosynthesis within the algae, or by feeding the algae sugar directly, and subsequent anaerobic fermentation of these carbon sources to produce ethanol under dark conditions. The ethanol is secreted into the culture media, and is collected in the headspace of the reactor and stored. This process may be drastically less capital and energy intensive. The process would essentially eliminate the need to separate the biomass from water and extract and process the oils (biodiesel) or starch (bioethanol).

In addition to ethanol, it is possible to use algae to produce other alcohols, such as methanol and butanol, using a similar process technology, although the recovery of heavier alcohols may be problematic and will need further R&D. The larger alcohols have energy densities closer to that of gasoline, but are not typically produced at the yields that are necessary for commercial viability.

2.3.2 Alkanes

In addition to alcohols, alkanes may be produced directly by heterotrophic metabolic pathways using algae; some of them produce a mix of hydrocarbons similar to light crude petroleum. These alkanes can theoretically be secreted and recovered directly, without the need for dewatering and extraction, and if desired, further treatment could be done to make a wide range of fuels.

According to Maxwell et al. (1985), for the implementation of an algae cultivation unit, a site selection and resource evaluation have to be performed considering several criteria (1) the water supply/demand, its salinity and chemistry;

(2) the land topography, geology and ownership; (3) the climatic conditions, temperature, insulation, evaporation, precipitation and (4) the easy access to nutrients and carbon supply sources.

In the production systems, it is essential to obtain a maximum biomass growth to very high cell densities. Cultivation conditions are complex with many inter-related factors, such as temperature, mixing, gas exchange, gas bubble size and distribution, light cycle and intensity, fluid dynamics and hydrodynamics stress, mineral and carbon regulation/bioavailability, cell fragility, cell density, water quality and salinity (Weissman et al. 1988; Barbosa et al. 2003a, b, 2004; Cho et al. 2007; Eriksen et al. 2007; Molina Grima et al. 1999; Perner-Nochta and Posten 2007; Kim et al. 2006; Gudin and Chaumont 1991; Ranga Rao et al. 2007a, b).

Finally, to produce biofuels at higher efficiencies, optimal media formulation is critical to ensure a sufficient and stable supply of nutrients to attain maximal growth acceleration and cell density (Schenk et al. 2008).

The design of photobioreactors is also a challenge to ensure good development of second-generation microalgal biofuels.

3 Microalgal Biomass Production

The systems for microalgal biomass production in large scale range from open ponds–shallow, circular tanks and raceways agitated from paddle wheel (Fig. 4) to a closed thin flat plate air lift photobioreactors (PBRs) (Fig. 5).

Fig. 4 Raceway pond agitated by paddle wheels

Fig. 5 Thin flat plate air lift photobioreactors (LNEG, Portugal)

Besides, these two systems others configurations were tried all over the world, namely column (Fig. 6) and tubular (vertical and horizontal) (Figs. 7, 8, respectively).

New developments seek to combine high productivity and low auxiliary energy demand with low cost criteria for large-scale application (Morweiser et al. 2010). Subitec (2010) (Stuttgart, Germany) presented an improved flat panel airlift design, with a total volume of 180 L represented a low-cost design, good mixing, utilization of the "flashing light effect" and short light paths without any unexposed zones (Fig. 9).

Flat panel (or flat plate) PBRs, supports the highest densities of photoautrotophic cells and promotes the highest photosynthetic efficiency (Rodolfi et al. 2009; Eriksen 2008).

Solix Biofuels (Fort Collins, CO, USA) (2010) have developed series of reactors surrounded by water, which act, as the same time, as a scaffold, temperature regulation and light diffuser (Fig. 10).

A very similar idea has been followed by Proviron (Belgium) (2010), where the PBR consists of thin vertical panels in one big translucent plastic bag (Fig. 11)

Both reactors, open and closed, present pros and constraints that are summarised in Table 6.

The PBR auxiliary energy demand should be as low as possible, and includes pumping, gassing, temperature regulation, nutrient supply, water and nutrient recycling, compounds extraction, refinement.

Temperature, nutrients and light, should be adequately controlled in the PBR-growing microalgae. To optimize them, light requirements of microalgae are one

Fig. 6 Vertical column photobioreactors (LNEG, Portugal)

of the most important parameter to be addressed, so that light will be provided at the appropriate intensity, duration and wavelength. Excessive intensity may lead to photo-oxidation and photo-inhibition, whereas low light levels will become growth limiting.

   Once the microalga strain has been chosen, considerations on the light type to be supplied (i.e. appropriate wavelengths) will be in order, so as to assure a high level of photosynthetic efficiency. Several selection criteria of artificial light sources for cultivation of photosynthetic microorganisms have been proposed; these include high electrical efficiency, low heat dissipation, reliability, durability,

Fig. 7 Tubular verticals photobioreactors

Fig. 8 Tubular horizontals photobioreactors

Fig. 9 Flat panel airlift reactor (Subitec) (http://www.subitec.com). Photo courtesy of Subitec

Fig. 10 New pilot reactor from Solix Biofuels: **a** low-ceiling design; **b** water as support (http://www.solixbiofuels.com). Photo courtesy of Solix Biofuels

Fig. 11 New pilot reactor from Proviron, Belgium (http://www.proviron.com). Photo courtesy of Proviron

long lifetime, compactness, low cost and a spectral output that falls within the absorption spectrum of the microorganism of interest (Bertling et al. 2006).

Full-spectrum light, about half of which is photosynthetically useful (400–700 nm), is normally used for microalgal growth; however, it has already been recognized that blue (420–450 nm) and red (660–700 nm) light are as efficient for photosynthesis as the full spectrum.

Recently, LEDs have been purposed, as a light source for PBR as they are cheap, have longer life expectancy, lower heat generation, high conversion efficiency, small enough to fit the reactor and a greater tolerance for switching on and off (Chen et al. 2011). In addition to LED, optical fibers (OF) excited by artificial lights are another potential source to improve microalgae culture systems, because it enhances the light conversion efficiency of the PBR, as they can provide uniform light distribution and can directly be immersed in the culture medium.

A great economically advantage should be the use of solar panels and wind power generator to supply all of the energy required by the multi-LED light sources. The conceptual photobioreactor combining OF (sunlight) and multi-LED light sources with solar panels and a wind power generator has the potential to be developed into a commercially viable microalgae cultivation system with zero electricity consumption (Chen et al. 2011).

Table 6 Main design features of Open and Closed Photobioreactors (adapted from Pulz 2001; Carvalho et al. 2006; Harun et al. 2010a)

Feature	Open systems	Closed systems
Cultivation		
Area-to-volume ratio	Large (4–10 times higher)	Small
Algal species	Restricted	Flexible
Species selection	Growth competition	Shear-resistance
Contamination	Possible	Unlikely
Cultivation period	Limited	Extended
Water loss through evaporation	Possible	Prevented
Controlling of growth conditions	Very difficult	Easy
Light utilization efficiency	Poor/fair	Fair/excellent[a]
Gas transfer	Poor	Fair/high
Temperature	Highly variable	Required cooling
Temperature control	None	Excellent
Automatic cooling system	None	Built in
Automatic heating system	None	Built in
Cleaning	None	Required due to wall growth and dirt
Microbiology safety	None	UV
Harvesting efficiency	Low	High
Biomass production		
Biomass quality	Variable	Reproducible
Biomass productivity	Low	High
Population density	Low	High
Operational mode		
Air pump	Built in	Built in
Shear	Low	High
CO_2 transfer rate	Poor	Excellent
Mixing efficiency	Poor	Excellent
Water loss	Very high	Low
O_2 concentration	Low due to continuous spontaneous out gassing	Exchange device
CO_2 loss	High	Low
Economics		
Land required	High	Low
Capital investment	Small	High
Periodical maintenance	Less	More
Operating cost	Lower	Higher
Harvesting cost	High	Lower
Most costly parameters	Mixing	O_2, Temp[a] control
Scale up technology for commercial level	Easy to scale up	Difficult in most PBR models

[a] Dependent on transparency of construction material

3.1 Hybrid Systems

Open ponds are a very efficient and cost-effective method of cultivating algae, but they could become contaminated with unwanted species very quickly. PBR are excellent for maintaining axenic cultures, but setup costs are generally ten times higher than for open ponds. A combination of both systems are probably the most logical choice for cost-effective cultivation of high yielding strains for biofuels. Inoculation has always been a part of algal aquaculture. Open ponds are inoculated with a desired strain that was invariably cultivated in a bioreactor, whether it was as simple as a plastic bag or a high-tech fibre optic bioreactor. Importantly, the size of the inoculum needs to be large enough for the desired species to establish in the open system before an unwanted species. Although sooner or later contaminating species will end up dominating an open system (if they do not required extreme conditions) and it will have to be cleaned and re-inoculated.

Therefore, to minimize contamination issues, cleaning or flushing the ponds should be part of the aquaculture routine, and as such, open ponds can be considered as batch cultures.

This process has been demonstrated by Aquasearch (Hawaii, USA) cultivating *Haematococcuspluvialis* for the production of astaxanthin. Half of the Aquasearch facility was devoted to PBR and half to open ponds. *H. pluvialis* is grown continuously in PBR under nutrient sufficient conditions and then a portion is transferred to nutrient-limited open ponds to induce astaxanthin production.

Enough nutrients are transferred with the inoculum for the culture to continue to grow for one day, and after 3 days when astaxanthin level peak, the open ponds are harvested, cleaned and then re-inoculated (Huntley and Redalje 2007). This approach is also very suitable for biofuel production, as under low-nutrient conditions algae rapidly start to convert energy from the sun into chemical energy stored as lipids as a means of survival (Gouveia et al. 2009).

The growing, harvesting and processing of any feedstock, including algal biomass requires considerable energy. The use of fossil-based energy sources for these actions would reduce the net carbon gain in a life-cycle assessment for this new fuel pathway (Subhadra and Edwards 2010). However, if algal energy production uses non-fossil, renewable energy sources, such as wind and solar energy, the process will show a substantial net carbon gain. One of the challenges in the algal biomass is the year round production of biomass, which is greatly dependent on the light and temperature. In colder months (3–5 months), outdoor algal growing facilities and photobioreactors need to be controlled for optimum algal growth, while in summer temperatures, the PBR need to be refreshed. A green house-based algal production may need heat to sustain high productivity in winter. Greenhouses with solar panels to harvest solar energy or greenhouses to operate with the heat from geothermal would substantially contribute to the sustainability issue.

4 Algal Cultivation

Besides the equipment needed for microalgae growth, it is essential to pay close attention to the selection of the most adequate species and strains, their cultivation conditions and nutrients available for their growth. In most cases, the production of any compound will rely on already available species and strains that have shown to be adequate due to their compound content and/or productivity. Typically, sources of microalgae, include existing collections of microalgae, commercially available either from Universities or other National and International Foundations or isolated from local waters and soils from diverse environments.

A multicriteria strategy has to be considered in the cultivation process (for microalgae and production units), and important factors should be attained such as

- growing rate, normally measured by total amount of biomass accumulated per unit time and unit volume
- nutrients availability, in particular of carbon dioxide sources when the goal of carbon sequestration is also deemed relevant;
- robustness and resistance to environmental conditions changes, such as nutrients, light, temperature and contamination from other microorganisms
- biomass harvesting and downstream processing

Therefore, it is crucial to understand how to select the right algae species, create an optimal photo-biological formula for each purpose, and build a cost-effective cultivation unit, no matter the size of the facility, or its geographical location.

Microalgae may assume many types of metabolisms (autotrophic, heterotrophic, mixotrophic and photo-heterotrophic) and are capable of a metabolic shift as a response to changes in the environmental conditions (Table 7).

Some organisms can grow (Chojnacka and Marquez-Rocha 2004):

- Photoautotrophically, i.e. using light as a sole energy source that is converted to chemical energy through photosynthetic reactions.
- Heterotrophically, i.e. utilizing only organic compounds as carbon and energy source.
- Mixotrophically, i.e. performing photosynthesis as the main energy source, though both organic compounds and CO_2 are essential. The organisms are able to live either autotrophically or heterotrophically, depending on the concentration of organic compounds and light intensity available.
- Photo-heterotrophycally, describes the metabolism in which light is required to use organic compounds as carbon source. The photo-heterotrophic and mixotrophic metabolisms are not well distinguished; in particular, they can be defined according to a difference in the energy source required to perform growth and specific metabolite production.

Although microalgae biomass production is strain dependent, heterotrophic growth could give much better productivity than other cultivation conditions. However, heterotrophic culture can get contaminated very easily, especially in

Table 7 Comparison of the characteristics of different cultivation conditions (Chen et al. 2011)

Cultivation condition	Energy source	Carbon source	Cell density	Reactor scale-up	Cost	Issues scale-up
Phototrophic	Light	Inorganic	Low	Open pond or PBR	Low	Low cell density High condensation cost
Heterotrophic	Organic	Organic	High	Conventional Fermentor	Medium	Contamination High subtract cost
Mixotrophic	Light and organic	Inorganic and organic	Medium	Closed PBR	High	Contamination High equipment and subtract cost
Photo-heterotrophic	Light	Organic	Medium	Closed PBR	High	Contamination High equipment and subtract cost

open cultivation systems, causing problems in large-scale production. In addition, the cost of an organic carbon source and the production of CO_2 is also a major concern from the commercial aspect. Phototrophic cultivation is most frequently used, despite a slow cell growth and low biomass productivity, because its are the easiest process and has lower cost to scale up, and because microalgae could uptake CO_2 from the flue gas of factories, which represent a notable advantage.

To date, there is not much information on the literature concerning using mixotrophic and photo-heterotrophic cultivation for microalgal production, but those two cultivation conditions are also restricted by contamination risk and light requirements, and may require the design of a special photobioreactor for scaling-up, thereby increasing the operation cost (Chen et al. 2011).

Not only the carbon source (inorganic or organic), nutrients (e.g. nitrogen and phosphorous) and vitamins, minerals are vital for algal growth, but also the equilibrium between operational parameters (temperature, light intensity and regime, dissolved oxygen, CO_2, pH, and product and byproduct removal) (Williams 2002).

The optimization of strain-specific cultivation conditions is of large complexity, with many interrelated factors that can be limiting. These include, light (cycle and intensity), temperature, nutrient concentration, O_2, CO_2, pH, salinity, water quality, mineral and carbon regulation/bioavailability, cell fragility, cell density and growth inhibition, mixing, fluid dynamics and hydrodynamic stress, depth, gas bubble size and distribution, gas exchange, mass transfer, dilution rate, toxic chemicals, presence of pathogens (bacteria, fungi, viruses) and competition by other algae and harvest frequency.

After light, the *Temperature* is the most important limiting factor for culturing algae in both closed and open outdoor systems. The temperature effects for many microalgae species in the laboratory are quite well established, however, the magnitude of temperature effects in the annual biomass production outdoors is not yet sufficiently acknowledged. Many microalgae can easily tolerate temperatures up to 15°C lower than their optimal, but exceeding the optimum temperature by only 2–4°C may result in the total culture loss (Moheimani and Borowitzka 2006), which represent a enormous problem in closed systems during the hot days, where the temperature inside the reactor may reach 55°C. In this case, evaporative water cooling systems may be economically used to decrease the temperature to around 20–26°C (Moheimani and Borowitzka 2006).

Salinity, in both open and closed systems, can affect the growth and cell composition of microalgae. Every alga has a different optimum salinity range that can increase during hot weather conditions due to high evaporation. Salinity changes normally affect phytoplankton in three ways:

1. osmotic stress
2. ion (salt) stress
3. changes in the cellular ionic ratios due to the membrane selective ion permeability (Moheimani and Borowitzka 2006).

The easiest way for salinity control is by adding fresh water or salt as required.

Mixing is another important growth parameter because it homogenizes the cells distribution, heat, metabolites, and facilitates gases transfer. In addition, a certain degree of turbulence, especially in large-scale production, is desirable to promote the fast circulation of microalgae cells from the dark to the light zone of the reactor (Barbosa 2003). However, high-liquid velocities and degrees of turbulence (due to mechanical mixing or air bubbles mixing) can damage microalgae due to shear stress (Eriksen 2008). The optimum level of turbulence is strain dependent and should be found for each alga to avoid decline in productivity (Barbosa 2003).

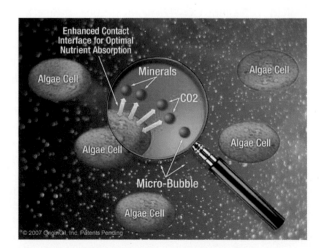

Fig. 12 Micro-bubbles to enhance nutrient and gases absorption (OriginOil 2010) (http://www.originoil.com). Photo courtesy of OriginOil Inc.

When the mixing is promoted by the introduction of bubbling air (air-lift-type reactors), it is favorable that the bubbles are small as possible to enhance the contact interface to maximize gas and nutrients transfer (Fig. 12).

Common biological contaminants observed include unwanted algae, mould, yeast, fungi, and bacteria. Attempts made to cultivate some microalgae species in raceway ponds failed, because cultures collapse due to the predation by protozoa and contamination by other algal species.

A way to decrease contaminants and improve yield is to subject the culture to a temporarily extreme change of the environmental factors, such as temperature, pH, or light, after removing the unwanted organism. This is particularly used in open systems, because in closed ones, the control over these parameters are much higher, in addition to the higher cell concentration.

5 CO_2 Sequestration

Atmospheric CO_2 levels have already exceeded 450 ppm CO_2-e (CO_2 equivalent) and are at levels classified as "dangerously high" (IPCC 2007; Stern 2006). Carbon dioxide is one of the major GHGs (greenhouse gas emissions) (Lopez et al. 2009) and the combustion of fossil fuels is the main source of CO_2, representing about 75% of the total anthropogenic emissions.

At present, reducing the use of fossil fuels or promoting CO_2 capture and sequestration seem to be the only way to cut or mitigate CO_2 emissions.

Although the development of CO_2-neutral biofuel production systems is important, their production will largely serve to stabilize atmospheric CO_2 levels at a "dangerously high" level (an important first step), rather than actively reducing it back down to an acceptable concentration (Schenk et al. 2008).

In light of this, reducing CO_2 emissions have become a popular research topic around the world. Various physical, chemical, and biological methods have been applied to capture CO_2 (Benemann 2003; Abu-Khader 2006). Physical sequestration of atmospheric CO_2 is often considered challenging, as it is technically difficult to separate CO_2 from other atmospheric gases.

Photosynthetic organisms have, however, fine-tuned this process over millions of years and of course are well adapted to capture CO_2 and store it as biomass. If this captured CO_2 could, therefore, be converted to a more stable form for long-term storage (~ 100 years or more), it would open up the important opportunity to couple CO_2-neutral biofuel production (e.g. biodiesel) with atmospheric CO_2 sequestration (Schenk et al. 2008)

Among the biological methods, the use of microalgae and cyanobacteria is considered one of the most effective approaches to fix CO_2 (de Morais and Costa 2007; Wang et al. 2008) and has attracted much attention in the last years because it leads to the production of biomass energy in the process of CO_2 fixation through photosynthesis. This in turn can contribute further to climate change mitigation by affecting the gas exchange of crops and soils (Marris 2006; Lehmann 2007).

Microalgae have the particularity to fix CO_2 efficiently from different sources, including atmosphere, industrial exhaust gases and soluble carbonate salts (e.g. $NaHCO_3$ and Na_2CO_3) (Wang et al. 2008).

In general, microalgae grow much faster than terrestrial plants, and the CO_2-fixation efficiency of microalgae is about 10–50 times faster than that of terrestrial plants (Wang et al. 2008).

Because the atmosphere contains only 0.03–0.06% CO_2, it is expected that mass transfer limitation could slow down the cell growth of microalgae (Chelf et al. 1993). On the other hand, industrial exhaust gases, such as flue gas contains up to 15% CO_2 (Maeda et al. 1995) providing a CO_2-rich source for microalgal cultivation and a potentially more efficient route for CO_2 biofixation.

Weissman and Tillett (1992) studied the capture of carbon dioxide by large pond-type systems. When operating under optimum conditions, the capture efficiency has been shown to be as high as 99% (Weissman and Tillett 1992; Zeiler et al. 1995). Based on the following equation, 1.57 g of CO_2 is required to produce 1 g of glucose:

$$6CO_2 + 12H_2O + light + chlorophyll \rightarrow C_6H_{12}O_6 + 6O_2 + 6H_2O$$

Kurano et al. (1995) reported fixation of 4 g CO_2/L day at growth rates of 2.5 g alga/L day, a ratio of 1.6 to 1. Taking into consideration the conversion of glucose into other compounds, such as lipids or starch under certain conditions the consumption of CO_2 can be as high as 2 g CO_2 to 1 g algae. Assuming a growth rate of 50 g/m^2 day, it is possible for 1 ha. of algal ponds to sequester up to one ton of CO_2 a day.

Flue gases from power plant are responsible for more than 7% of the total world CO_2 emissions. Because CO_2 in flue gas is available at not much or no cost means that production costs of algal biomass could be about 15% lower (Doucha et al. 2005).

Several microalgae species can tolerate high CO_2 concentration in the gas stream (Table 8) (and high temperatures) and moderate levels of SO_x and NO_x (up to 150 ppm) (Matsumoto et al. 2003). It is important to notice that the optimum flue gas injection rate into the photobioreactor is dependent on the time course of irradiance and culture temperature (Doucha et al. 2005).

Chlorococcum littorale, a marine alga, showed exceptional tolerance to high CO_2 concentration of up to 40% (Iwasaki et al. 1998; Murakami and Ikenouchi 1997).

de Morais and Costa (2007) reported the microalgae *S. obliquus*, *Chlorella kessleri* and *Spirulina* sp. as also exhibited good tolerance to high CO_2 contents (up to 18% CO_2) indicating their great potentials for CO_2 fixation from CO_2-rich streams.

For *Spirulina* sp., the maximum specific growth rate and maximum productivity were 0.44/day and 0.22 g/L day, with both 6 and 12% CO_2 (v/v), respectively, while the maximum cell concentration was 3.50 g/L (dry basis) with both CO_2

Table 8 Some microalgae strains studied for CO_2 bio-sequestration (adapted from Wang et al. 2008)

Microalga	CO_2 (%)	Temperature	Biomass productivity (g/L day)	CO_2 fixation rate (L day)
Chlorococcum littorale	40	30	N/A	1.0
Chlorella kessleri	18	30	0.087	0.163[a]
Chlorella sp. UK001	15	35	N/A	>1
Chlorella vulgaris	15	–	N/A	0.624
Chlorella vulgaris	Air	25	0.040	0.075[a]
Chlorella vulgaris	Air	25	0.024	0.045[a]
Chlorella sp.	40	42	N/A	1.0
Dunaliella	3	27	0.17	0.313[a]
Haematococcus pluvialis	16–34	20	0.076	0.143
Scenedesmus obliquus	Air	–	0.009	0.016
Scenedesmus obliquus	Air	–	0.016	0.031
Botryococcus braunii	–	25–30	1.1	>1.0
Scenedesmus obliquus	18	30	0.14	0.26
Spirulina sp.	12	30	0.22	0.413[a]

[a] Calculated from the biomass productivity according to equation, CO_2 fixation rate = 1.88 × biomass productivity, which is derived from the typical molecular formula of microalgal biomass, $CO_{0.48} H_{1.83}N_{0.11}P_{0.01}$ (Chisti 2007)

concentrations. For *S. obliquus*, the corresponding maximum growth rate and maximum productivity were 0.22/day and 0.14 g/L day, respectively.

Murakami and Ikenouchi (1997) by an extensive *screening*, selected more than ten strains of microalgae with high capability of fixing CO_2. Two green algal strains, *Chlorella* sp. UK001 and *Chlorococcum littorale*, showed high CO_2 fixation rates exceeding 1 g CO_2/L day. *Botryococcus braunii* SI-30, which showed the ability of producing high content of hydrocarbons, was recommended as a promising candidate for combined CO_2 mitigation and biofuel production (Murakami and Ikenouchi 1997).

Ho et al. (2010) reported a CO_2 consumption rate of 549.90 mg/L day for a maximum *S.obliquus* biomass productivity and lipid productivity of 292.50 mg/L day and 78.73 mg/L day (38.9% lipid content per dry weight of biomass), respectively, in cultivated two-stage system with 10% CO_2.

The pH has an important influence on the CO_2 absorption capacity-since it can be enhanced in alkaline conditions (Hsue et al. 2007). However, the pH increase due to photosynthetic functions under batch cultivation and growth would be inhibited via alkali or amounts of useful carbon source limitations (Shiraiwa et al. 1993). Commonly, different pH levels would have variations in the ratios of $CO_2/HCO_3^-/CO_3^{2-}$. Those variations might change the useful carbon sources due to biological physiology related to carbon-type transporters (Price et al. 2004). Therefore, different ratios of C(useful)/N would happen under the same amount of carbon source addition, but at different pHs.

In consequence, those variations might change the biochemical compositions via different pathways under carbon or nitrogen limitations (Zhila et al. 2005). Usually, the variations of macromolecular compositions were focused on lipids, proteins, nucleic acids, and carbohydrates (Brown et al. 1996a, b; Stehfest et al. 2005).

6 Microalgal Biomass Harvesting

The harvesting process is energy dependent and represent the main percentage of the total production costs (20–30%) being still considered as a major limiting factor (Molina Grima et al. 2003).

Algae cultures typically have high water contents and to remove the large quantities of water and process large algal biomass volumes, a suitable harvesting method should be attained. It may involve one or more steps and be achieved in several physical, chemical, or biological ways, to perform the desired solid–liquid separation. Experience has demonstrated that albeit a universal harvesting method does not exist, this is still an active area for research, being possible to develop an appropriate and economical harvesting system for any algal species.

The first challenge is to concentrate cells from relatively dilute solutions of ca 0.5–5 g/L dry weight to solutions between 20 and 100% more concentrated than the starting solution.

Most common harvesting methods, include gravity sedimentation, centrifugation, filtration and microscreening, ultra-filtration, flotation, sometimes with an additional flocculation step or with a combination of flocculation–flotation, and electrophoresis techniques (Uduman et al. 2010). The cost of algae harvesting can be high, because the mass fractions in culture broth are generally low, while the cells normally carry negative charge and excess algogenic organic matters (AOM) to keep their stability in a dispersed state (Danquah et al. 2009).

The selection of harvesting technique is dependent on the properties of microalgae, such as density, size and value of the desired products (Brennan and Owende 2010).

For example, the cyanobacterium *Spirulina*'s long spiral shape naturally lends itself to the relatively cost- and energy-efficient microscreen harvesting method (Benemann and Oswald 1996).

Microalgae harvesting can generally be divided into a two-step process, including:

- Bulk harvesting. The purpose of this is to separate microalgal biomass from the bulk suspension. By this method, the total solid mater can reach 2–7% using flocculation, flotation, or gravity sedimentation (Brennan and Owende 2010).
- Thickening. The purpose of this harvesting is to concentrate the slurry, with filtration and centrifugation usually applied in this process. This step needs more energy than bulk harvesting (Brennan and Owende 2010).

6.1 Gravity Sedimentation

Sedimentation is commonly applied for separating microalgae in water and waste-water treatment. Density and radius of algae cells and the induced sedimentation velocity influence the settling characteristic of suspended solids (Brennan and Owende 2010). Although sedimentation is a simple process, it is very slow (0.1–2.6 m/h) (Choi et al. 2006) and in high-temperature environments, the bio-mass could be deteriorated. Enhanced microalgal harvesting by sedimentation can be achieved through lamella separators and sedimentation tanks (Uduman et al. 2010). The success of solids removal by gravity settling depends highly on the density of microalgal particles. Edzwald (1993) found that low-density microalgal particle do not settle well, and are unsuccessfully separated by settling.

Flocculation is frequently used to increase the efficiency of gravity sedimentation.

6.2 Centrifugation

Sedimentation and centrifugation can be described by Stokes' law, which predicts that its velocity is proportional to the difference in density between the cell and medium on the one hand and on the square of the radius of the cells (Stokes radius) on the other hand. Although for bacteria gravitational force-based methods are not easy to apply, for yeast and microalgae with diameters >5 μm and relatively thick cell walls they are feasible. Most microalgae can be recovered from the liquid broth using centrifugation. Laboratory centrifugation tests were conducted on pond effluent at 500–1,000g and showed that about 80–90% microalgae can be recovered within 2–5 min (Molina Grima et al. 2003). Pure sedimentation—or settling as it is called in aquaculture—is employed in some algal farms, but is time- and space-consuming and is not an appropriate choice for biodiesel production. Knuckey et al. (2006) also states that the exposure of microalgal cells to high gravitational and shear forces can damage cell structure. According to Molina Grima et al. (2003), centrifugation is a preferred method, especially for producing extended shelf-life concentrates for aquaculture, however, they agree that this method is time-consuming and costly. Energy costs of about 1 kW/h m^3 have been quoted for centrifugation.

Commercial centrifuges accelerating to at least 10,000×g enhance separation and decanting centrifuges have also been successfully employed (e.g. Westfalia 2010). Currently, centrifugation is considered to be too costly and energy intensive for the primary harvesting of microalgae. The energy input alone has been esti-mated at 3,000 kW/ton (Benemann and Oswald 1996). Centrifugation is however a very useful secondary harvesting method to concentrate an initial slurry (10–20 g/L) to an algal paste (100–200 g/L) and could possibly be used in com-bination with oil extraction (Schenk et al. 2008).

In addition, these devices can be easily cleaned or sterilized to effectively avoid bacterial contamination or fouling of raw product.

Heasman et al. (2000) reported 88–100% cell viability and around 95–100% harvesting efficiency by centrifugation at $13,000 \times g$, although it is not cost-effective due to high-power consumption, especially when considering large volumes.

6.3 Flocculation

Microalgae carry a negative charge that prevents them from self-aggregation within suspension. The surface charge on the algae can be countered by the addition of chemicals known as flocculants. Flocculation is a process in which dispersed particles are aggregated together to form large particles for settling. The increased particle size leads, therefore, to faster sedimentation (see Stokes' law) or better, interaction with flotation bubbles.

These cationic chemicals coagulate the algae without affecting the composition and toxicity of the product. Types of flocculants can include, inorganic, such as $Al_2(SO_4)_3$ (aluminium sulphate), $FeCl_3$ (ferric chloride) and $Fe_2(SO_4)_3$ (ferric sulphate) or organic. These multivalent salts are commonly used and vary in effectiveness, which is directly related to the ionic charge of the flocculant. Knuckey et al. (2006) used Fe^{3+} flocs with induced pH to harvest various kinds of algae and achieved the efficiencies around 80%.

Nevertheless, the addition of flocculants is currently not a method of choice for cheap and sustainable production. Recent developments involve encouraging self-flocculation of the cells, which can occur during carbon limitation or pH shifts.

6.4 Autoflocculation

Certain species naturally flocculate, while others flocculate in response to environmental stimuli, nitrogen stress, pH and level of dissolved oxygen. Autoflocculation occurs as a result of precipitation of carbonate salts with algal cells in elevated pH, a consequence of photosynthetic CO_2 consumption with algae (Sukenik and Shelef 1984). Hence, prolonged cultivation under sunlight with limited CO_2 supply assists autoflocculation of algal cells for harvesting. Laboratory experiments also revealed that autoflocculation can be simulated by the addition of NaOH to achieve certain pH values.

6.5 Chemical Coagulation

Adding chemicals to microalgal culture to induce flocculation is a common practice in various solid–liquid separation processes as a pre-treatment stage, which is applicable to the treatment of large quantities of numerous kinds of microalgal species (Lee et al. 1998). There are two main classifications of flocculants according to their chemical compositions: (1) inorganic flocculants and (2) organic flocculants/polyelectrolyte flocculants. The utilization of microorganisms to recover microalgae has also been investigated, by studying the use of *Paenibacillus* sp. for effective harvesting of microalgae (Oh et al. 2001).

6.5.1 Inorganic Coagulants

Microalgal cells are negatively charged, as a result of adsorption of ions originating from organic matter and dissociation or ionization of surface functional groups (Uduman et al. 2010). By disrupting the stability of the system, successful microalgal harvesting can be obtained. Addition of a coagulant, like iron-based or aluminum-based coagulants, will neutralize or reduce the surface charge (Molina Grima et al. 2003). Alum was utilized for harvesting of *Scenedesmus* and *Chlorella* via charge neutralization (Molina Grima et al. 2003). Microalgae can also be flocculated by inorganic flocculants at sufficiently low pH (Uduman et al. 2010). However, despite its advantages, coagulation using inorganic coagulants suffers from the following drawbacks:

- a large concentration of inorganic flocculant is needed to cause solid–liquid separation of the microalgae, thereby producing a large quantity of sludge.
- the process is highly sensitive to pH level.
- although some coagulants may work for some microalgal species, they do not work for others.
- the end product is contaminated by the added aluminum or iron salts.

6.5.2 Organic Flocculants

To achieve effective sedimentation, floc size should be more than 100 μm, with the addition of a high-molecular weight bridging polymer increasing floc size and improving microalgal settling (Edzwald 1993).

Flocculation by aluminum sulfate followed by certain polyelectrolytes is effective in microalgal harvesting (Pushparaj et al. 1993). Biodegradable organic flocculants, such as chitosan, are produced from natural sources that do not contaminate the microalgal biomass (Divakaran and Pillai 2002), being the biomass able to be used in food, feed and nutraceuticals. The most effective flocculants for the recovery of microalgae are cationic flocculants (Bilanovic et al. 1988). Anionic and nonionic polyelectrolytes have been shown to fail to flocculate microalgae, which is explained by the repulsion existing between charges or the insufficient distance to bridge particles. Polymer molecular weight, charge density of molecules, dosage, concentration of microalgal biomass, ionic strength and pH of the broth, and the extent of mixing in the fluid have all been found to affect floccu-lation efficiency (Molina Grima et al. 2003).

Bilanovic et al. (1988) noted that flocculation by cationic polymers can be inhibited by the high salinity of a marine environment. High-molecular weight polyelectrolytes are generally better bridging agents. A high biomass concentra-tion in the broth also helps flocculation due to the frequent cell–cell encounters. Mixing at a low level is thus useful, as it helps in bringing the cells together, but excessive shear forces can disrupt flocs. In addition to all of the factors mentioned before, functional groups on microalgal cell walls are important, because they stimulate the formation of negative charge centers on the cell surfaces (Uduman et al. 2010).

6.6 Combined Flocculation

A combined flocculation process is a multistep process using more than one type of flocculant. Sukenik et al. (1988) studied a combined flocculation process with marine microalgae. To induce flocculation in sea water, two methods were found. The first is combining polyelectrolytes with inorganic flocculants, such as ferric chloride or alum, and the second is ozone oxidation followed by flocculant addition. Vandamme et al. (2010) demonstrated the feasibility of using cationic starch for flocculation of both fresh and marine water microalgae.

Another interesting method recently presented by Massingill is to feed the algae to the fish Tilapia (*O. mosambicus*) which obtain very little nutrient from it. The algal biomass is then harvested from the sedimented droppings by a conveyor belt (10–14% solids) and then air-dried (Carlberg et al. 2002).

6.7 Filtration and Screening

Although filtration is often applied at a laboratory scale, in large-scale applications it suffers from problems, such as membrane clogging, the formation of com-pressible filter cakes and in particular, from high maintenance costs. Cost-effective filtration is limited to filamentous or large colonial microalgae. The cost of applying tangential flow filtration relies on membrane replacement and pumping, and large-scale harvesting using this method is limited by this. Pressure or vacuum filtration can be used to recover relatively large microalgae, but concentration of the microalgae is required for these processes to be effective. Power consumptions for these operations are of the order of 0.3–2 kW/h m^3, not dissimilar to those required for centrifugation (Molina Grima et al. 2003).

Microstrainer and vibrating screen filters are two of the primary screening devices in microalgae harvesting and are attractive methods because of their mechanical simplicity and availability in large unit sizes. Microstrainers can be realized as rotating filters of a very fine mesh screens with frequent backwash. Microstrainers have several advantages, such as simplicity in function and construction, easy operation, low investment, negligible abrasion as a result of absence of quickly moving parts, being energy intensive and having high filtration ratios. Nevertheless, a high microalgal concentration can result in blocking the screen, whereas a low microalgal concentration or when applied to organisms approaching bacterial dimensions can result in inefficient capture (Wilde et al. 1991). The Molina Grima et al. (2003) study confirmed this result and concluded that it would be necessary to flocculate the cells before microstraining.

Tangential flow filtration is a high rate method for microalgal harvesting and recovery of 70–89% of freshwater algae could be possible (Petrusevski et al. 1995). In addition, tangential flow filtration retains the structure, properties and motility of the collected microalgae.

Although the successful laboratory studies for concentrating microalgae, used in downstream fractionation (Rossignol et al. 1999; Rossi et al. 2004), a definitive study on large-scale algal harvesting is yet to be published. Lazarova et al. (2006) work has shown that the cost of microfiltering river water can be as low as 0.2 kW h/m^3 of water processed. Decreasing the process volume by at least a factor of 100, significantly lowers the costs of disruption and fractionation stages downstream.

Several variables associated with the choice of membranes and type of organisms could increase this cost, and there is a considerable scope for optimization of this process. As a guide to potential improvement, the costs of desalination by reverse osmosis, where a far higher pressure process is used, have fallen dramatically (85%) over the past decade to give a total production cost of about $1 m^{-3} and with desalination energy costs being as low as 3 kW/h m^3. This is largely down to a better membrane technology, greater membrane longevity, increased scale of operation and better system management and such advances might also be expected in membrane separation processes for harvesting of microalgae (Greenwell et al. 2010).

6.8 Flotation

Flotation is a gravity separation process in which air or gas bubbles are attached to solid particles and then carry them to the liquid surface.

Flotation is a commonly used approach to remove microalgae from reservoir water prior to its use as drinking water. Typically, the water is initially ozonated, after which the sensitized cells are then treated with about 10 ppm polyelectrolyte salts (typically salts of aluminium and iron or formulations of charged organic polymers) prior to being subjected to flotation. Based on the bubble sizes used in the flotation process, the applications can be divided into dissolved air flotation (DAF), dispersed flotation and electrolytic flotation.

Chen et al. (1998) noted that flotation is more beneficial and effective than sedimentation with regard to removing microalgae.

6.8.1 Dissolved Air Flotation

Dissolved air flotation (DAF) involves the generation fine bubbles produced by a decompression of pressurized fluid. The fine bubbles less than 10 mm adhere to the flocs making them very buoyant and causing them to increase rapidly to the surface of a separation tank (Uduman et al. 2010). Flotation can capture particles with a diameter of <500 μm by collision between a bubble and a particle and the subsequent adhesion of the bubble and the particle (Yoon and Luttrell 1989).

The resultant concentrated cell foam (7–10% dry weight) is then removed as slurry.

These processes work well in fresh water and are capable of dealing with the large volumes required in a commercial scale plant (greater than 10,000 m^3/day) (Crossley et al. 2002), where additions of ozone and flocculant are made. The main disadvantage of this approach is the contamination of the materials with the floc agent, which may significantly decrease their value (Molina Grima et al. 2003).

Factors determining DAF harvesting of microalgae, include the pressure of the tank, recycle rate, hydraulic retention time, and floating rate of particle. Chemical flocculation has been used with DAF to separate microalgae (Uduman et al. 2010). Microalgae autoflocculation using dissolved oxygen which is produced photosynthetically has also been studied after flocculation using alum or C-31 polymer (Koopman and Lincoln 1983), and about 80–90% microalgal removal was obtained when about 16 mg/L microalgal float concentration was used. Edzwald (1993) found that DAF removed microalgae more effectively than settling, although flocculation pre-treatment was required in the former process.

6.8.2 Dispersed Air Flotation

Dispersed air flotation entails 700–1,500 μm bubbles formed by a high-speed mechanical agitator with an air injection system (Rubio et al. 2002). Chen et al. (1998) compared dispersed air flotation efficiencies for microalgae using three collectors, and noted that the cationic N-cetyl-N-N-N-trimethylammonium bromide (CTAB) effectively removed Scenedesmusquadricauda, while the nonionic X-100 and anionic sodium dodecylsulfate did not. They attributed these differences to changes in surface hydrophobicity with collector adsorption.

6.9 Electrolytic Separation

The electrolytic method is another potential approach to separate algae without the need to add any chemicals. In this method, an electric field drives charged algae to move out of the solution (Mollah et al. 2004). Water electrolysis generates hydrogen that adheres to the microalgal flocs and carries them to the surface.

Electro-coagulation mechanisms involve three consecutive stages:

- generation of coagulants by electrolytic oxidation of the sacrificial electrode
- destabilization of particulate suspension and breaking of emulsion
- aggregation of the destabilized phases to form flocs.

Azarian et al. (2007) investigated the removal of microalgae from industrial waste-water using continuous flow electro-coagulation. Different from electrolytic coagulation, electrolytic flocculation does not requires the use of sacrificial electrodes. Electrolytic flocculation works based on the movement of microalgae to the anode to neutralize the carried charge and then form aggregates. Poelman et al. (1997) showed that the efficiency of algal removal is 80–95% when electrolytic flocculation is applied.

There are several benefits to use electrochemical methods, including environmental compatibility, versatility, energy efficiency, safety, selectivity, and cost effectiveness (Mollah et al. 2004). An investigation into the removal of microalgae electrolytically in batch and continuous reactors by flotation was conducted by Alfafara et al. (2002). The results for a batch system showed that by increasing the electrical power input, the rate of chlorophyll removal increased and the electrolysis time decreased.

Gao et al. (2010a, b) studied the algae removal by electro-coagulation–flotation (ECF) technology and indicated that aluminum was an excellent electrode material for algae removal when compared with iron. The optimal parameters determined were current density $= 1$ mA/cm^2, pH $= 4$–7, water temperature $= 18$–36°C, algae density $= 0.55 \times 10^9$–1.55×10^9 cells/L. Under the optimal conditions, 100% of algae removal was achieved with the energy consumption as low as 0.4 kW/m^3. The ECF performed well in acid and neutral conditions. At low initial pH of 4–7, the cell density of algae was effectively removed in the ECF, mainly through the charge neutralization mechanism; while the algae removal worsened when the pH increased (7–10), and the main mechanism shifted to sweeping flocculation and enmeshment. Furthermore, initial cell density and water temperature could also influence the algae removal. Overall, the results indicated that the ECF technology was effective for algae removal, from both the technical and economical points of view (Gao et al. 2010a, b).

Recently, OriginOil company is employing several next-generation technologies to greatly enhance algae cultivation and oil extraction (OriginOil 2010), by going on to control the harvesting and oil extraction cycles in a high-speed, round-the-clock, streamlined industrial production of algae oil.

In the process, mature algae culture is injected through the OriginOil device, where Quantum Fracturing^{TM}, pulsed electromagnetic fields and pH modification (using CO_2) combine to break the cell walls, thereby releasing the oil within the cells (Fig. 13).

The processed culture now travels into a settling tank, or gravity clarifier (Fig. 14), to fully separate into oil, water and biomass. Algae oil increases to the

Fig. 13 Harvesting lipids in the Live Extraction process (OriginOil 2010) (http://www.originoil.com). Photo courtesy of OriginOil Inc.

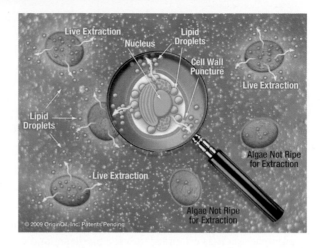

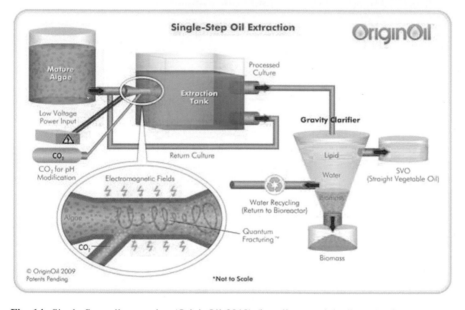

Fig. 14 Single-Step oil extraction (OriginOil 2010) (http://www.originoil.com). Photo courtesy of OriginOil Inc.

top for skimming and refining, while the remaining biomass settles to the bottom for further processing as fuel and other valuable products (OriginOil 2010).

Although there are several biomass harvesting methods, Richmond (2004) suggested one main criterion for selecting a proper harvesting procedure is the desired product quality. In one hand for low value products, gravity sedimentation may be used, possibly enhanced by flocculation. Sedimentation tanks or settling ponds are also possible, e.g. to recover biomass from sewage-based processes.

On the other hand for high-value products, to recover high-quality algae, such as for food, feed and nutraceuticals, it is often recommended to use continuously operating centrifuges that can process large volumes of biomass.

Albeit at considerable cost, centrifuges are suitable to rapidly concentrate any type of microorganisms, which remain fully contained during recovery. Another basic criterion for selecting the harvesting procedure is its potential to adjust the density or the acceptable level of moisture in the resulting concentrate right to the optimum subsequent process (Molina Grima et al. 2003; Richmond 2004).

7 Cell Disruption

Cell disruption is often necessary for recovering intracellular products from microalgae, such as oil and starch for biodiesel and ethanol production, as well as added value compounds. To open the cell, various ways were tested, including, e.g., freezing, alkalic and organic solvents, osmotic shocks, sonication, high-pressure homogenization, and bead milling (Chisti and Moo-young 1986; Molina Grima et al. 2004).

The most important processing parameters for disintegration, which can be controlled directly, are: feed rate of the suspension, agitator speed, cell density, bead diameter, bead density, bead filling (% of the grinding chamber volume), geometry of the grinding chamber, and design of the stirrer (Kula and Schütte 1987; Engler 1993; Hatti-Kaul and Mattiasson 2003).

Industrially relevant are horizontal bead mills (Kula and Schütte 1987; Middelberg 1995) originally designed for the homogenization and size reduction of different commercial products such as milk and paint.

7.1 Bead Mill Homogenizers

The principle of bead mills is based on the rapid stirring of a thickened suspension of microorganisms in the presence of beads. The disruption occurs by the crushing action of the glass beads, as they collide with the cells. In bead milling, a large number of minute glass or ceramic beads are vigorously agitated by shaking or stirring (Hopkins 1991). The basic setup of a bead mill is a jacketed grinding chamber with a rotating shaft through its centre. The shaft is fitted with discs that impact kinetic energy to small beads in the chamber, forcing them to collide with each other. The beads are retained in a grinding chamber by a sieve or an axial slot smaller than the bead size. The beads are accelerated in a radial direction, forming stream layers of different velocity and creating high-shear forces. An external pump feeds the suspension into the grinding chamber.

When compared with high-pressure methods of cell disruption wet bead milling is low in shearing forces. Membranes and intracellular organelles can often be isolated intact. The method has been used for years to disrupt microorganisms. It is considered the method of choice for disruption for spores, yeast and fungi and works successfully with tough-to disrupt cells such as cyanobacteria, mycobacteria, spores and microalgae (Hopkins 1991).

The size of the beads is important. Optimal size for bacteria and spores is 0.1, 0.5 mm for yeast, mycelia, microalgae, and unicellular animal cells, such as leucocytes or trypsinized tissue culture cells and 1.0 or 2.5 mm for tissues, such as brain, muscle, leaves and skin. Speed of disruption is increased about 50% by using like-sized ceramic beads made of zirconia silica or zirconia rather than glass (Hopkins, unpublished observations), presumably because of their greater density. Really tough tissue sometimes require chrome-steel beads—which are five times more dense than glass beads. Generally, the higher the volume ratio of beads to cell suspension, the faster the rate of cell disruption. After treatment, the beads settle by gravity in seconds and the cell extract is easily removed by pipette.

There are more bead mills types, such as shaking type, rotor type and rotor types, however, with less application on the microalgae cell disruption.

7.2 Freeze Fracturing

Both microbial pastes and plant and animal tissue can be frozen in liquid nitrogen and then ground with a common mortar and pestle at the same low temperature. Presumably, the hard frozen cells are fractured under the mortar because of their brittle nature. In addition, ice crystals at these low temperatures may act as an abrasive.

A freeze-fracturing device look like a tablet press, the pulverizer consists of a hole machined into a stainless steel base into which fits a piston. The base and piston are pre-cooled to liquid nitrogen temperatures. The hard frozen animal or plant tissue is placed in the hole. The piston is placed in the hole and given a sharp blow with a hammer. The resulting frozen, powder-like material can be further processed by other methods.

7.3 Ultrasonic Disintegrators

Are widely used to disrupt cells. These devices generate intense sonic pressure waves in liquid media. Under the right conditions, the pressure waves cause formation of microbubbles which grow and collapse violently. Called cavitation, the implosion generates a shock wave with enough energy to break cell membranes and even break covalent bonds.

Modern ultrasonic processors use piezoelectric generators made of lead zirconate titanate crystals. The vibrations are transmitted down a titanium metal horn or probe tuned to make the processor unit resonate at 15–25 kHz. The rated power output of ultrasonic processors vary from 10 to 375 W. What really counts are the power density at the probe tip. Higher output power is required to sustain good performance in large-sized probes. For cell disruption, probe densities should be at least 100 W/cm^2 and the larger the better for tip amplitude (typical range 30–250 μm).

Ultrasonic disintegrators generate considerable heat during processing. For this reason, the sample should be kept ice-cold. For microorganisms, the addition of 0.1–0.5 mm diameter glass beads in a ratio of one volume beads to two volumes liquid is recommended, although this modification will eventually erode the sonicator tip (Hopkins 1991).

Cerón et al. (2008) to extract lutein from the microalga *Scenedesmus almeriensis* tested three cell disruption methods, such as mortar (125 mL volume), bead mill (2 L volume and a rotation speed of 120 rpm, with ceramic beads of 28 mm diameter), and ultrasound (Pselecta Ultrasons unit) and combination between them. Their results demonstrate that cell disruption is necessary, and that the best option among the treatments tested with regard to industrial applications was the use of bead mill with alumina in a 1:1 w/w as disintegrating agent for 5 min.

Converti et al. (2009) point out the use of ultrasounds (mod. UP100H, Hielscher, Teltow, Germany) combined with chloroform/methanol allowed the complete extraction of the microalgae fatty components.

OriginOil (2010) company claimed for an efficient and continuous live extraction without destroying the algae cell. The process can be operated in parallel with Cascading Production^{TM} to create a combined cycle with increased productivity. Live extraction, or 'milking', is inherently efficient because the single algae cell can produce more oil during its lifetime using lower amounts of energy.

8 Production Rates, Production and Processing Costs

Unfortunately, there is a lack of data in the public domain on the production rates as well as on production and processing costs, due to industry withholding research results. Biofuel production also reportedly requires biomass at a cost of less than $300 US/ton dry weight. Ben-Amotz presented open pond yields averaging 20 g/m^2.day and that overall production costs of $340 US/ton are viable if the lipid content is high enough (Schenk et al. 2008). Nevertheless, van Beilen (2010) presented the status as it is today for microalgae yield (5–60 ton/ha year) and production costs ($5.000–$15.000/ton), both for cultivation and harvesting.

van Beilen (2010) referred a cost of $15/m^2 for open systems and ten times more for close ones, being the cost of algal biomass of $8.000–15.000/ton and $30.000–70.000/ton, for the open and close systems, respectively.

Cheap bioreactors designs were presented by Bryan Willson using disposable plastic materials and Ben Cloud at a cost ~$15 US/m^2 that have standard farm-style set ups (Schenk et al. 2008).

Proviron (2010) claimed for an investment below $25/m^2, with a low power consumption (less than 20 kW/ha) and a high cell density (10 g/L). The price for closed systems is much higher ($1.1 M/ha against $125,000/ha for the open systems) (Chisti 2009). On the other way, Subitec (2010) claimed a capital costs $175,000,000 (25% for photobioreactors). Carlsson et al. (2007) also estimated a algae production cost of $210/ton but state, however, that even with the most favourable assumptions with this low costs and high revenues for fuels ($125/ton algae) and GHG ($65/ton), the process would still not be economically feasible.

The vast bulk of microalgae cultivation today is growth in open ponds. These systems can be built and operated very economically. The main disadvantage of the open systems is that they need big land area extension; they loose water by evaporation and they are also susceptible to contamination by unwanted species. The contamination drawback can limit this cultivation system to algal strains which can only grow under severe conditions (ex *Spirulina* (*Arthrosphira*), high alkalinity; *Dunaliella*, high salinity).

The typical theoretical productivity is around 0.025 kg/m^2 day (82 ton/ha year) and the maximum biomass concentration 1 g/L (Chisti 2009). However, these productivities are no longer seen as realistic and typical biomass yields of commercial systems are in the range of 10–30 ton/ha year (van Beilen 2010).

Closed PBR (tubular, plate or bubble column) have some advantages; namely, the large surface-to-volume ratio, the better control of algae culture and gas transfer, the lower evaporation and contamination, and higher biomass productivity [1.535 kg/m^2 day ($\sim$158 ton/ha year)] and higher algae cell densities are possible (4 g/L) (Chisti 2009). Subitec (2010) stated an annual productivity of 120 ton/ha. The commercial bioreactor supplier Algae Link claim year-round productivity of several species of algae in the order of 365 ton/ha year for one of their systems (Singh et al. 2011). On the other end, Green Fuel Technologies Corporation (USA), who have several large-scale pilot plants operating focus on CO_2 capture from industrial emitters, indicates productivities of $\sim$250–300 ton/ha year (Singh et al. 2011).

Ono and Cuello (2006) also doubt on the economic feasibility and declare that to achieve the target CO_2 mitigation price of $30/ton CO_2 at 40% biological conversion efficiency, the allowable net cost should be less than $2.52/m^2 year at low-light intensity (average US location).

Gallagher (2011) summarized the data from other authors concerning algae farm costs and productivities, and after normalization to 2009 dollars, reached to the following average values:

Productivity: 100,000 ton/ha year
Lipid concentration: 35%/wt biomass
Biodiesel yield: 39.500 L/ha
Capital costs: $112.400/ha
Fuel operating costs: $39.300/ha

Although the economic feasibility of algal-to-biofuel seems to be fair and dependent on government subsidies and the future price of oil, in addition to optimized biomass yields (Gallagher 2011), the requirement of carbon-neutral renewable alternatives makes microalgae one of the best future sources of biofuels (Chisti 2007). The process is technically feasible, the potential for CO_2 sequestration is high, and for it to be cost-effective and energy-efficient economically viable, good harvesting methods, de-watering, supply of CO_2 and downstream processing are required (Mata et al. 2010).

The cost of CO_2, for instance, has strong influence on profitability, and could decrease significantly, if carbon taxes are imposed in high-carbon emitting industries (e.g. power plants) (Gallagher 2011).

According to Chisti (2007), for algal diesel to potentially replace fossil fuels, it must be priced as follow:

$$C_{\text{algaloil(per L)}} \leq 6.9 \times 10^{-3} \times C_{\text{petroleum(per L)}}$$

The cost of crude oil would have to exceed \$100 per barrel to make a high return scenario plausible. Despite many economists feel that this is unlikely to occur in the near-term period (3–5 years) due to present economic conditions and the low global demand for oil, the geologists and petroleum engineers are predicting that global oil production rates will soon peak and then begin to decline, resulting in a steady increase in oil prices (Gallagher 2011).

It is widely accepted that microalgal biomass could assist in fossil fuels in the near future, if commercial production will intertwine the following requisites;

- highly productive microalgae that could be cultivated on a large scale using wastewater as a nutrient supply, and waste CO_2, as carbon source
- harvesting, dewatering and extraction of algal biomass could be developed at a low cost
- production of biofuels could be combined with that of higher value co-products (Figs. 15, 16, 17).

If these issues are resolved following long-term R&D and concerted efforts by the public and private sectors in addition to large investments in the area, microalgae cultures might become an economically viable, renewable and carbon-neutral source of transportation biofuels, which do not jeopardize our forests and food supply.

The much higher productivity of microalgae cultures and the absence of competition for arable land and water resources justify the long-term R&D required.

Furthermore, fuel from algae represents a market that is worth hundreds of billions of dollars (Table 9).

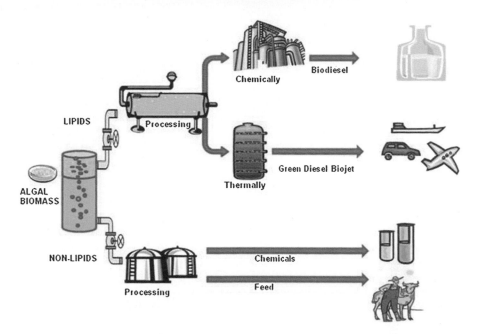

Fig. 15 The potential of microalgal biomass: processing and production of fuels and co-products

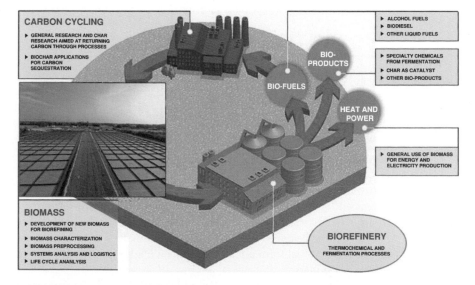

Fig. 16 Biorefinery concept of microalgae biomass (© PetroAlgae LLC www.petroalgae.com). Photo courtesy of PetroAlgae LLC

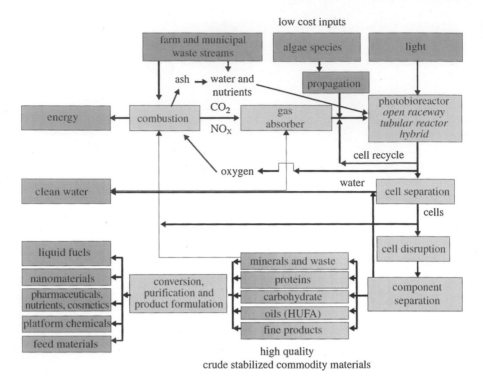

Fig. 17 Schematic of a microalgae biomass biorefinery concept based upon the production of several products from waste materials allowing their complete utilization, such as liquid fuels, commodity chemicals and materials for high-value formulated products (Greenwell et al. 2010)

Table 9 Biofuel potential in 2014 (billion gallons) (Oilgae 2009)

Total oil consumption in 2014	1,500
Total projected supply by traditional biofuels	41
Total ethanol production in 2014	26
Total biodiesel production in 2014	15
Share of traditional biofuels in total oil consumption	2.73%
Projected market size for traditional biofuels	$123 billion

Assumption: one gallon of oil = $3

From the above table, it is clear that even by 2014, less than 3% of total fossil fuels will be replaced by biofuels from traditional sources. Even this small percentage represents a market share of over $100 billion. Algae have the potential to replace a much higher percentage of fossil transportation fuel than traditional feedstock. This implies that fuel from algae represents a market share that is worth hundreds of billions of dollars.

According to Lee (2011), microalgae will undoubtedly become an important feedstock for diesel in the future. This author, adopting the Taiwan General Equilibrium Model Energy for Biofuels (TAIGEM-EB), estimated for 2040 that the share of petroleum of the total production energy will reduce to 19.24%, while algal biodiesel will reach 19.24%. CO_2 emissions will further have a reduction of 21.7%, representing an impressive result. This study attempts to intensify the effort of incorporating advanced technology and this scenario will only be possible if provided by a strong government support, such as subsidies or tax breaks (Gallagher 2011) from a developed economy country (Lee 2011).

9 Life Cycle Analysis

Sustainability is a key principle in natural resource management, and it involves operational efficiency, minimization of environmental impact and socio-economic considerations; all of which are interdependent (Brennan and Owende 2010).

Life cycle analysis is a widely accepted method of quantifying the environmental impacts of products. Life cycle analysis (LCA) appears as a relevant tool to evaluate new technologies for bioenergy production. This tool identifies the technological bottlenecks and therefore supports the ecodesign of an efficient and sustainable production chain.

Carbon and nitrogen emissions (GHG), other nutrients, emissions (P,K), water and land uses, net energy balance, eutrophication potential, impact on biodiversity, soil erosion and pesticide use are generally evaluated.

Besides some feedstocks currently used for bioenergy, such as corn, grain sorghum, oil palm, perennial grasses (miscanthus, switchgrass), rapeseed, short rotation trees (birch, poplar, willow), soybean, sugar beet, sugarcane, and sweet sorghum, the algae biomass offers a clear lowest land intensity impact (Miller 2010; Clarens et al. 2010), contrarily to soybean, which reaches the worst position.

Grain sorghum, rapeseed and soybean ranked the lowest for both land use and nitrogen intensity (Miller 2010). Sugarcane is always the most highly ranked. It should be noted that these results take into account only the feedstock as a whole, without including the co-product that eventually could be produced. Reijnders (2009) also pointed the fact that microalgae do not appear to outperform terrestrial plants such as sugarcane, when both the conversion of solar energy into biomass and the life cycle inputs of fossil fuels are considered.

Algae are nitrogen intensive and require significant additional nitrogen inputs for small increases in energy yield (Clarens et al. 2010; Miller 2010). Lardon et al. (2009) confirm the potential of microalgae as an energy source, but highlight the imperative necessity of decreasing the energy and fertilizer consumption. Nevertheless, this constraint could be avoided, with recycling nitrogen by wastewater treatment. To demonstrate the benefits of algae production coupled with wastewater treatment, Clarens et al. (2010) included in their model three different municipal wastewater effluents. Each provided a significant reduction in the environmental burdens of algae cultivation, and the use of urine was found to make algae even more environmentally beneficial. According to Chinnasamy et al. (2010a, b) wastewater generated by carpet mills along with sewage from the

Dalton area in north-central Georgia (40–55 million m^3/year) has the potential to generate up to 15,000 ton of algal biomass, which can produce about 2.5–4 million liters of biodiesel and remove about 1,500 ton of nitrogen and 150 tons of phosphoros from the wastewater in 1 year.

The water footprint and nutrients balance studies of biodiesel production using microalgae, by Yang et al. (2011), confirmed the competitiveness of microalgae-based biofuels and highlighted the necessity of recycling harvested water and using sea/wastewater as water source, in agreement with Singh et al. (2011). To generate 1 kg of biodiesel, 3,726 kg water, 0.33 kg nitrogen and 0.71 kg phosphate are required if freshwater is used without recycling. Recycling harvest water reduces the water and nutrient usage by 84 and 55%, respectively. Using sea/wastewater decreases 90% water requirement and eliminates the need of all the nutrients except phosphate (Yang et al. 2011).

Eutrophication potential is the other impact in which algae perform favorably in comparison to terrestrial crops, while energy consumption, water use and GHG are higher, according to Clarens et al. (2010). However, Campbell et al. (2011) indicate a reduction of GHG emissions and costs, using microalgae with high annual growth rates and favorable soil conditions as in Australia.

Collet et al. (2011) performed the life-cycle assessment of biogas production from the microalgae C. vulgaris and compared the results from the algal biodiesel with the first-generation biodiesels. They highlighted the productivity of algae per hectare and per year, its ability to recycle CO_2 from flue gas and the capacity of the algae to carry out anaerobic digestion directly, to produce methane and recycle nutrients (N, P and K). Their results suggested that the impacts generated by the production of methane from microalgae strongly correlated with the electric consumption, and progresses could be achieved by decreasing the mixing costs and circulation between different production steps, or by improving the efficiency of the anaerobic process under controlled conditions (Collet et al. 2011).

The important work carried out by Stephenson et al. (2010) pointed out that if the future target for the productivity of lipids from microalgae, such as C. vulgaris, of ~ 40 tons/ha year could be achieved, cultivation in typical raceways of depth ~ 0.3 m would be significantly more environmentally sustainable than fossil-derived diesel and many first-generation biofuels, as well as than in closed air-lift tubular bioreactors. While biodiesel produced from microalgae cultivated in raceway ponds would have a GWP $\sim 80\%$ lower than fossil-derived diesel (on the basis of the net energy content), if airlift tubular bioreactors were used, the GWP of the biodiesel would be significantly greater than the energetically equivalent amount of fossil-derived diesel.

The electricity required during cultivation was found to contribute the most to the overall requirement for fossil energy and GWP of the biodiesel produced from C. vulgaris cultivated in either raceways or air-lift tubular bioreactors. In contrast, the fossil energy requirements and GWP of each of the algal processing steps were found to be significantly lower than that for the cultivation, and the burdens associated with the transport of the oil feedstock and biodiesel product were negligible.

The GWP and fossil energy requirement in this operation were found to be particularly sensitive to (1) the yield of oil achieved during cultivation, (2) the velocity of circulation of the algae in the cultivation facility, (3) whether the culture media could be recycled or not, and (4) the concentration of carbon dioxide in the flue gas.

The use of life cycle assessment is of crucial importance to guide the future development of biodiesel from microalgae. Moreover, there is an urgent need for pilot-scale trials of algal biodiesel production to allow LCA of actual operations (Stephenson et al. 2010).

10 Modelling Approaches

Mathematical approaches (computer models, simulations) may be used to enhance the development and exploitation of microalgae for commercial gain (Greenwell et al. 2010). Modeling techniques are important in the:

- optimization of algal growth and production of specific end products (Flynn 2001, 2003, 2008a, b; Chi et al. 2007);
- optimization of bioreactor design and operation (Camacho Rubio et al. 2003; Mulbry et al. 2008a, b; Cooney et al. 2011);
- production facility operation (Jones et al. 2002; Hu et al. 2008a, b; Cooney et al. 2011); and
- coupled operation and financial modeling and risk analysis (Lee 2011; Gallagher 2011).

11 Algal Biorefinery Strategy

The term biorefinery was coined to describe the production of a wide range of chemicals and biofuels from biomasses through the integration of bioprocessing and appropriate low environmental impact chemical technologies in a cost-effective and environmentally sustainable manner (Li et al. 2008).

The microalgal biomass biorefinery concept is not new; however, it assisted in making biofuel production economically viable. An algal biorefinery could potentially integrate several different conversion technologies to produce biofuels including biodiesel, green diesel, green gasoline, aviation fuel, ethanol, and methane, as well as valuable co-products, such as fats, polyunsaturated fatty acids, oil natural dyes, sugars, pigments (mainly β-carotene and astaxanthin), antioxidants and polyunsaturated fatty acids (EPA, DHA).

Conceptually, the biorefinery would involve sequentially the cultivation of microalgae in a microalgal farming facility (CO_2 mitigation), extracting bioreactive products from harvested algal biomass, thermal processing (pyrolysis, liquefaction or gasification), extracting high-value chemicals from the resulting liquid, vapor and/or solid phases, and reforming/upgrading biofuels for different applications (Li et al. 2008).

After oil and/or starch removal from the microalgal biomass (for biodiesel and/or ethanol production, respectively), the leftover biomass can be processed into methane or livestock feed, used as organic fertilizer due to its high N:P ratio, or simply burned for energy cogeneration (electricity and heat) (Wang et al. 2008).

The high-value bioactive compounds could be used in nutritional supplements, food/feed additives, aquaculture, cosmetics, pharmaceuticals, biofertilizers, edible vaccines through genetic recombination (Chisti 2006; Rosenberg et al. 2008) and pollution prevention. Microalgae play an imperative role in bioremediation and wastewater treatment. They can eliminate heavy metals, uranium and other pollutants from wastewater, and they can degrade carcinogenic polyaromatic hydrocarbons and other organics. Furthermore, algae are accountable for at least 50% of the photosynthetic biomass production in our planet and they are great sources of biofuels (Chisti 2006).

Nevertheless, several authors (e.g. van Harmelen and Oonk 2006; van Beilen 2010; Park et al. 2011), point that only if the algal biomass is a by-product of wastewater treatment systems, GHG abatement and/or of the production of high-value compounds, such as astaxanthin, β-carotene, EPA and DHA, commercially viable energy production from algal biomass might be feasible.

Mussgnug et al. (2010) proposed another approach for the microalgae producers of H_2. The cells, response to the induction of H_2 production cycle is the strong increase of starch and lipids (high fermentative potential compounds), which results in an increase in the biogas production (second step), after a first step of H_2 production, showing a synergistic effect in a biorefinery concept.

A microalgae biomass biorefinery proposed by © PetroAlgae company is represented in Fig. 16, whereas the one projected by Greenwell et al. (2010) is shown in Fig. 17 where little or no waste products are present and allows for residual energy capture, recycling of unused nutrients, and water purification and recycling. This system would mean low environmental impact and maximization of the value of products from the system (Greenwell et al. 2010).

Subhadra and Edwards (2010) stated that is would be better to integrate a renewable energy park (IREP) where the facilities are centralized, instead of a single central facility, such as giant petroleum refineries operated by a single firm. Major firms can be a part of IREPs and might play an important role in the development of this concept. However, other small-scale renewable energy (wind, solar, geothermal and biomass) firms, working as a consortium, may also be an integral component of IREPs. Together, these firms can cross-feed power, heat, raw materials and products with the shared goal of minimizing emissions to the atmosphere and optimizing the utilization of natural resources such as land, water and fossil fuels, and fossil agricultural chemicals.

The integration of established prototype carbon capture devices, which feed algal cultures, should also be examined (Fig. 18). Several novel green technologies such as geothermal heat pumps (Dickinson et al. 2009), dual fuel (bivalent) ground source heat pumps (Ozgener and Hepbasil 2007), solar-assisted heat pump systems

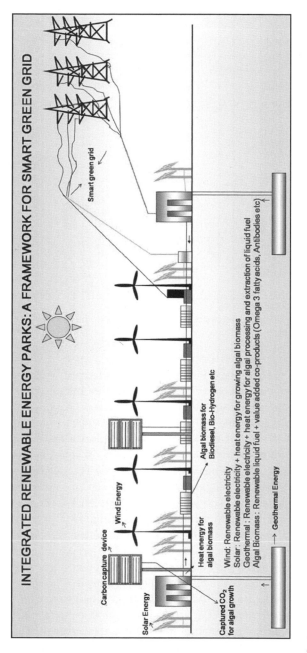

Fig. 18 Integrated renewable energy parks: a frame work for a "smart green grid" with net zero carbon emission (Subhadra and Edwards 2010)

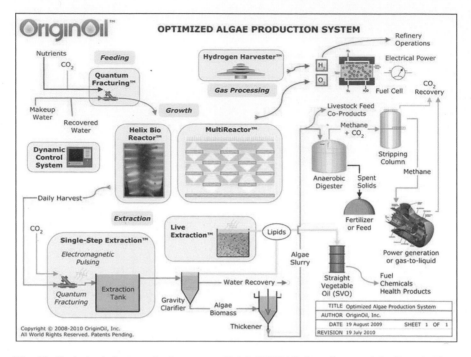

Fig. 19 Optimized algae production system (OriginOil 2010) (http://www.originoil.com). Photo courtesy of OriginOil Inc

(Benli and Durmus 2009), solar wind turbine (which harvest wind and sun energy in one element) have been receiving increased attention because of their potential to reduce primary energy consumption and thus reduce GHG emission. Further, newer energy conservation and utilization concepts such as bioheat from wood (Ohlrogge et al. 2009), bioelectricity from biomass (deB Richter et al. 2009) and hybrid hydrogen–carbon process for the production of liquid hydrocarbon fuels (Agrawal et al. 2007) can also be envisioned into the broader design concept of IREPs. Together, these technologies and concepts can maximize the ecological and environmental benefits of energy production from IREPs. The green electricity from these IREPs may flow into the existing grid (Fig. 18).

OriginOil (2010) proposed an integrated system called *Optimized Algae Production System* (Fig. 19) where the first step is a low-pressure Quantum Fracturing^{TM}. It works by breaking up carbon dioxide and other nutrients into micron-sized bubbles and infusing them into the growth vessel. The growth occurs in OriginOil's Helix BioReactor^{TM}, which features a rotating vertical shaft of low energy lights (highly-efficient LEDs) arranged in a helix or spiral pattern are tuned precisely to the waves and frequencies for optimal algae growth. They claim that in the Cascading Production^{TM} and Single-Step Extraction^{TM}, the oil and biomass are separat without having to dewater the algae, and the continuous process is called Live Extraction^{TM}. After extraction, the water is recycled back into the

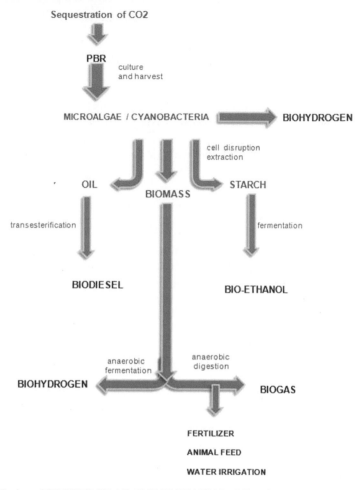

Fig. 20 Project PTDC/PTDC/AAC-AMB/100354/2008: Microalgae as a sustainable raw material for biofuel production (biodiesel, bioethanol, bio-H_2 and biogas). A biorefinary concept: integration of different processes such as biohydrogen production from microalgae and/or cyanobacteria. Oil and starch extraction from microalgae biomass, to biodiesel and bioethanol production, respectively; from the left-over biomass, biohydrogen and biogas production, by anaerobic fermentation and anaerobic digestion, respectively

system. The harvested oil is packaged for refining and distribution, and the algae mass is devoted to various 'green' applications such as fuel, animal feed, fertilizers, chemicals, health products and construction materials (OriginOil 2010).

The author team developed a project microalgae as a sustainable raw material for biofuels production (biodiesel, bioethanol, bio-H_2 and biogas) (PTDC/PTDC/AAC-AMB/100354/2008), where it was proposed as an integrated system (using a biorefinery concept). The optimization of all energy vectors (biodiesel, bioethanol, biohydrogen and biogas) will be highlighted (Fig. 20).

The best microalgae for CO_2 mitigation will be selected (from a hospital residues incinerator) and screening of microalgae that synthesize oil and/or starch (to biodiesel and/or bioethanol production, respectively) will be performed to select the best oil and/or starch producers. The microalgae and/or cyanobacteria that produce biohydrogen only need to be purified to be used in the industry or in fuel cells.

Downstream processing, mainly harvesting and cellular rupture, will be optimized for each microalga to ensure the best results.

The microalgae as a whole and the biomass leftovers after oil and/or starch extraction allow the production of biohydrogen and biogas by fermentation and anaerobic digestion, respectively, using different microorganisms and conditions. Anaerobic digestion waste can be used as fertilizer, water irrigation and/or animal feed. This integrated system of total biomass bioconversion makes the process more attractive and advantageous, from the technological and economical point of view.

Another interesting approach to the economy of the global process of microalgae is the concept of microbial fuel cells (MFC) where the algae photobioreactors (acting as cathodic half cells) can be coupled with another microbial organism (acting as anodic half cells). An example is the integrated bioethanol–biodiesel–microbial fuel cell facility, utilizing yeast and photosynthetic alga, as described by Powell and Hill (2009). Into an existing bioethanol production facility were integrated photosynthetic microalgae MFCs that capture CO_2 and generate electrical power and oil for biodiesel production. So, the facility would produce not only bioethanol, but also part of the power energy necessary for the bioethanol production process and oil to be sold for biodiesel production, with the reduction of CO_2 emissions from the bioethanol production process. The remaining biomass after oil removal can be sold as an animal feed supplement as well as carbon credits (Powell and Hill 2009).

Besides biorefinery and microbial fuel cells, the future of overproducing biofuels from microalgae is intertwined with metabolic engineering through genetic modification metabolic engineering (Chisti 2010).

The following issues represent a big challenge in terms of genetic and metabolic engineering but will enhance the overall performance of biofuel production from microalgal biomass (Chisti 2010; Singh et al. 2011): Cloning and transforming genes that influence the synthesis of lipids, starch or hydrogen, increasing photon conversion efficiencies by reducing the light-harvesting antenna complexes and improving robustness to severe environment; enhance biomass growth rate; increase oil content in biomass; improve temperature tolerance to reduce the expense of cooling; eliminate the light saturation phenomenon, making increased growth in response to increasing light level; reduce photo-inhibition; reduce susceptibility to photo oxidation that damages cells.

A report by Edwards (2010) was mostly in agreement with the author's opinion, related to a query on people linked with algae-based research (50%), algae producers (25%) and academics (20%). As much as 25% of the respondents had more than 5 years and 15% had 11–30 years of experience in the algae industry. This people were from over 50 countries and many from Europe, Asia, US, the Pacific Rim, India and the Middle East. The query essentially revealed the following conclusions: most industry participants believe that algal production will focus on three biofuels, such as biodiesel, ethanol and JP-8 jet fuel. Most growers plan to use algal strains from algal collections, genetically modified organisms or carefully selected strains from natural settings. Production models seem to vary based on the production objectives, type of feedstock and location. However, algal producers are experimenting with a diverse set of production models, using closed systems (33%) and open ponds (33%), and another 33% using semi-closed ponds or polycultures. About 35% indicated that algae should be cultivated in the tropics and 30% selected the mid-latitudes. Roughly 25% indicated that algae would be produced all over the earth (Edwards 2010).

The industry's most critical production challenges are production systems, harvest, extraction and component separation, followed by algal species selection, culture stability, quality control monitoring and contamination issues. Critical input challenges were dominated by total cost, followed by genetically modified organisms, water source and cost, land source and cost and nutrient source and cost. Life-cycle analysis and indirect landuse are also industry challenges.

Besides biofuels, the main co-products from algal biomass are likely to be organic fertilizer, animal feed, feed for fish and fowl, wastewater remediation and carbon capture and sequestration. Over 40% of respondents predicted that co-products would include chemicals and unique compounds, nutraceuticals, pharmaceuticals and food ingredients.

The social and economic issues that algae may address include carbon capture, cleaning polluted water, animal and fish feed, health foods and nutraceuticals, liquid transportation fuels, nutrient recovery, organic fertilizers, displacing oil imports and medicines and pharmaceuticals (Edwards 2010).

12   Conclusions

Algal biofuels have a tremendous potential for contributing to environmental, social and economic sustainability. Algal biofuel production should integrate other environmentally sustainable technologies, such as CO_2 sequestration, emissions cleanup from industrial and agricultural wastes and the purification of water and should be done in conjunction with the production of valuable co-products.

Many biofuels can be produced from "green coal", such as liquid fuels (e.g. biodiesel, green diesel, jet fuel and bioethanol) and gas fuels (e.g. biogas, syngas, and bio-hydrogen) which can be used in engines and turbines and as feedstock for refineries.

However, several drawbacks of microalgal biomass production should be solved, such as more efficient production, harvesting, dewatering, drying and extraction (if applicable).

Thus, fuel-only algal systems are not plausible, at least not in the foreseeable future, and additional revenues are required. Microalgal production should be assisted by alternative energies, for mixing the culture, illumination, dewatering and processing, and when needed for drying.

Both thermochemical liquefaction and pyrolysis appear to be feasible methods for the conversion of algal biomass to biofuels, after the extraction of oils from algae. Anaerobic digestion shows potential to reduce external energy demand and to recycle a part of the mineral fertilizers, avoiding eutrophication, especially when coupled with a wastewater treatment system. Bioethanol production from algae through fermentation could be another interesting alternative, due to the fact that it requires less energy consumption and is a simplified process when compared to biodiesel production.

To make algae-to-energy systems a practical reality, considerable research should continue such as genetic improvements, biorefinery and microbial fuel cell concepts, and the integration of alternative energies into wastewater and CO_2 treatment systems.

Acknowledgments The author thanks Eng° João Sousa (LNEG, Portugal) for the figures and Dr João Miranda and Dr[a] Ana Evangelista Marques (LNEG, Portugal) for their contributions to the Bioethanol and Biohydrogen sections, respectively. The author also thanks a native English speaker and friend, Stephanie Seddon-Brown, and the Eng[a] Ana Paula Batista for assistance in the manuscript correction. The work was supported by FCT Project PTDC/ENR/68457/2006.

References

Abu-Khader M (2006) Recent progress in CO_2 capture/sequestration: a review. Energy Sources 28:1261–1279

Agrawal R, Singh NR, Ribeiro FH, Delgass WN (2007) Sustainable fuel for the transportation sector. Proc Natl Acad Sci U S A 104:4828–4833

Alfafara CG, Nakano K, Nomura N, Igarashi T, Matsumura M (2002) Operating and scale-up factors for the electrolytic removal of algae from eutrophied lake water. J Chem Technol Biot 77:871–876

Al-Widyan MI, Al-Shyoukh AO (2002) Experimental evaluation of the transesterification of waste palm oil into biodiesel. Bioresour Technol 85:253–256

Amin S (2009) Review on biofuel oil and gas production process from microalgae. Energy Convers Manag 50:1834–1840

Angenent LT, Karim K, Al-Dahhan MH, Wrenn BA, Rosa D-E (2004) Production of bioenergy and biochemicals from industrial and agricultural wastewater. Trends Biotechnol 22:477–485

Antolin G, Tinaut FV, Briceno Y, Castano V, Perez C, Ramirez AI (2002) Optimization of biodiesel production by sunflower oil transesterification. Bioresour Technol 83:111–114

Azarian GH, Mesdaghinia AR, Vaezi F, Nabizadeh R, Nematollahi D (2007) Algae removal by electro-coagulation process, application for treatment of the effluent from an industrial wastewater treatment plant. Iran J Public Health 36:57–64

Balat M (2005) Current alternative engine fuels. Energy Sources 27:569–577

Balat M (2009) Possible methods for hydrogen production. Energy Sources 21:39–50

Barbosa MJGV (2003) Microalgal photobioreactors: scale-up and optimisation. Ph.D. thesis. Wageningen University, The Netherlands

Barbosa B, Jansen M, Ham N (2003a) Microalgae cultivation in air-lift reactors: modeling biomass yield and growth rate as a function of mixing frequency. Biotechnol Bioeng 82:170–179

Barbosa B, Albrecht M, Wijffels R (2003b) Hydrodynamic stress and lethal events in sparged microalgae cultures. Biotechnol Bioeng 83:112–120

Barbosa B, Hadiyanto M, Wijffels R (2004) Overcoming shear stress of microalgae cultures in sparged photobioreactors. Biotechnol Bioeng 85:78–85

BBC (2009) First flight of algae-fuelled jet. Available at http://news.bbc.co.uk/2/hi/science/nature

Becker EW (1994) Oil production. In: Carey NH, Higgins IJ, Potter WG (eds) Sir J Baddiley. biotechnology and microbiology. Cambridge University Press, Cambridge

Benemann JR (2003) Biofixation of CO_2 and greenhouse gas abatement with microalgae-technology roadmap. Final report submitted to the US Department of Energy, National Energy Technology Laboratory

Benemann J, Oswald W (1996) Systems and economic analysis of microalgae ponds for conversion of CO_2 to biomass. Final report to the US Department of Energy. Pittsburgh Energy Technology Center

Benli H, Durmus A (2009) Evaluation of ground-source heat pump combined latent heat storage system performance in greenhouse heating. Energy Build 41:220–228

Bertling K, Hurse TJ, Kappler U, Rakic AD (2006) Lasers—an effective artificial source of radiation for the cultivation of anoxygenic photosynthetic bacteria. Biotechnol Bioeng 94:337–345

Bilanovic D, Shelef G, Sukenik A (1988) Flocculation of microalgae with cationic polymers—effects of medium salinity. Biomass 17:65–76

Biller P, Ross AB (2011) Potential yields and properties of oil from the hydrothermal liquefaction of microalgae with different biochemical content. Bioresour Technol 102:215–225

Boichenko VA, Hoffmann P (1994) Photosynthetic hydrogen-production in prokaryotes and eukaryotes—occurrence, mechanism, and functions. Photosynthetica 30:527–552

Boichenko VA, Greenbaum E, Seibert M (2004) Hydrogen production by photosynthetic microorganisms. In: MDA, Barber J (eds) Photoconversion of solar energy, molecular to global photosynthesis, vol 2. Imperial College Press, London, pp 397–452

BP statistics (2009) http://www.bp.com/statisticalreview

Brennan L, Owende P (2010) Biofuels from microalgae—a review of technologies for production, processing, and extractions of biofuels and co-products. Renew Sust Energ Rev 14:557–577

Bridgwater AV, Meier D, Radlein D (1999) An overview of fast pyrolysis of biomass. Org Geochem 30:1479–1493

Brown LR (2006) Beyond the oil peak. In: Brown LR (ed) Plan B 2.0 rescuing a planet under stress and a civilization in trouble. W.W. Norton & Co., New York, pp 21–40

Brown MR, Dunstan GA, Norwood SJ, Miller KA (1996a) Effects of harvest stage and light on the biochemical composition of the diatom *Thalassiosira pseudonana*. J Phycol 32:64–73

Brown MR, Barrett SM, Volkman JK, Nearhos SP, Nell JA, Allan GL (1996b) Biochemical composition of new yeasts and bacteria evaluated as food for bivalve aquaculture. Aquaculture 143:341–360

Camacho Rubio F, Garcia Camacho F, Fernandez Sevilla JM, Chisti Y, Molina Grima E (2003) A mechanistic model of photosynthesis in microalgae. Biotechnol Bioeng 81:459–473

Campbell CJ (1997) The coming oil crisis. Multi-science Publishing Company and Petrocon-sultants, S.A Essex

Campbell PK, Beer T, Batten D (2011) Life cycle assessment of biodiesel production from microalgae in ponds. Bioresour Technol 102:50–56

Cantrell KB, Ducey T, Ro KS, Hunt PG (2008) Livestock waste-to-bioenergy generation opportunities. Bioresour Technol 99:7941–7953

Carlberg JM, van Olst JC, Massingill MJ, Chamberlain RJ (2002) Aquaculture wastewater treatment system and method of making same. Kent Seatech: US Patent 6,447,681, 10 Sept 2002

Carlsson AS, van Beilen JB, Moller R, Clayton D (2007) Micro- and macro-algae utility for industrial applications. In: Dianna B (ed) Outputs from the EPOBIO project. CPL press, UK

Carvalho AP, Meireles LA, Malcata FX (2006) Microalgal reactors: a review of enclosed system designs and performances. Biotechnol Prog 22:1490–1506

Cerón MC, Campos I, Sánchez JF, Acién FG, Molina Grima E, Fernandez-Sevilla JM (2008) Recovery of lutein from microalgae biomass: development of a process for *Scenedesmus almeriensis* biomass. J Agric Food Chem 56:11761–11766

Chelf P, Brown LM, Wyman CE (1993) Aquatic biomass resources and carbon dioxide trapping. Biomass Bioenergy 4:175–183

Chen YM, Liu JC, Ju YH (1998) Flotation removal of algae from water. Colloid Surf B 12:49–55

Chen P, Min M, Chen Y, Wang L, Li Y, Chen Q, Wang C, Wan Y, Wang X, Cheng Y, Deng S, Hennessy K, Lin X, Liu Y, Wang Y, Martinez B, Ruan R (2009) Review of the biological and engineering aspects of algae to fuels approach. Int J Agric Biol Eng 2:1–30

Chen C-Y, Yeh Ki L, Aisyah R, Lee D-J, Chang J-S (2011) Cultivation, photobioreactor design and harvesting of microalgae for biodiesel production: a critical review. Bioresour Technol 102:71–81

Cheryl (2010) Algae becoming the new biofuel of choice. Available online 2008. http://duelingfuels.com/biofuels/non-food-biofuels/algae-biofuel.php#more-115N

Chi ZY, Pyle D, Wen ZY, Frear C, Chen SL (2007) A laboratory study of producing docosahexaenoic acid from biodiesel-waste glycerol by microalgal fermentation. Process Biochem 42:1537–1545

Chiaramonti D, Oasmaa A, Solantausta Y (2007) Power generation using fast pyrolysis liquids from biomass. Renew Sust Energ Rev 11:1056–1086

Chinnasamy S, Bhatnagar A, Claxton R, Das K (2010a) Biomass and bioenergy production potential of microalgae consortium in open and closed bioreactors using untreated carpet industry effluent as growth medium. Bioresour Technol 101:6751–6760

Chinnasamy S, Bhatnagar A, Hunt RW, Das K (2010b) Microalgae cultivation in a wastewater dominated by carpet mill effluents for biofuel applications. Bioresour Technol 101:3097–3105

Chisti Y (2006) Microalgae as sustainable cell factories. Environ Eng Manag J 53:261–274

Chisti Y (2007) Biodiesel from microalgae. Biotechnol Adv 25:294–306

Chisti Y (2008a) Biodiesel from microalgae beats bioethanol. Trends Biotechnol 26:126–131

Chisti Y (2008b) Response to Reijnders: do biofuels from microalgae beat biofuels from terrestrial plants. Trends Biotechnol 26:351–352

Chisti Y (2009) Biodiesel from microalgae. Seminario Internacional de Biocombustibles de Algas. Antofagasta, Chile, 7–8 October

Chisti Y (2010) Fuels from microalgae. Biofuels 1:233–235

Chisti Y, Moo-young M (1986) Disruption of microbial cells for intracellular products. Enzyme Microb Technol 8:194–204

Cho S, Ji SC, Hur S, Bae J, Park IS, Song YC (2007) Optimum temperature and salinity conditions for growth of green algae *Chlorella ellipsoidea* and *Nannochloris oculata*. Fish Sci 73:1050–1056

Choi SK, Lee JY, Kwon DY, Cho KJ (2006) Settling characteristics of problem algae in the water treatment process. Water Sci Technol 53:113–119

Chojnacka K, Marquez-Rocha FJ (2004) Kinetic and Stoichiometric relationships of the energy and carbon metabolism in the culture of microalgae. Biotechnology 3:21–34

Clarens AF, Resurreccion E, White M, Colosi A (2010) Environmental life cycle comparison of algae to other bioenergy feedstocks. Environ Sci Technol 44:1813–1819

Clark J, Deswarte F (2008) Introduction to chemicals from biomass, Wiley Series in Renewable Resources, ISBN978-0-470-05805

Collet P, Hélias-Arnaud A, Lardon L, Ras M, Goy RA, Steyer JP (2011) Life-cycle assessment of microalgae culture coupled to biogas production. Bioresour Technol 102:207–214

Converti A, Casazza AA, Ortiz EY, Perego P, Del Borghi M (2009) Effect of temperature and nitrogen concentration on the growth and lipid content of *Nannochloropsis oculata* and *Chlorella vulgaris* for biodiesel production. Chem Eng Process 48:1146–1151

Cooney MJ, Young G, Pte R (2011) Bio-oil from photosynthetic microalgae: case study. Bioresour Technol 102:166–177

Costa JAV, Morais MG (2011) The role of biochemical engineering in the production of biofuels from microalgae. Bioresour Technol 102:2–9

Crossley IA, Valade MT, Shawcross J (2002) Using the lesson learned and advanced methods to design a 1500 Ml/day DAF water treatment plant. Water Sci Technol 43:35–41

Czernik S, Bridgwater AV (2004) Overview of applications of biomass fast pyrolysis oil. Energy Fuels 18:590–598

Danquah MK, Ang L, Uduman N, Moheimani N, Fordea GM (2009) Dewatering of microalgal culture for biodiesel production: exploring polymer flocculation and tangential flow filtration. J Chem Technol Biot 84:1078–1083

Das D (2009) Advances in biological hydrogen production processes: an approach towards commercialization. Int J Hydrogen Energ 34:7349–7357

de Morais MG, Costa JAV (2007) Biofixation of carbon dioxide by *Spirulina* sp and *Scenedesmus obliquus* cultivated in a three-stage serial tubular photobioreactor. J Biotechnol 129:439–445

deB Richter JrD, Jenkins JH, Karakash JT, Knight J, McCreery LR, Nemestothy KP (2009) Wood energy in America. Science 323:1432–1433

Dermibas A (2006) Oily products from mosses and algae via pyrolysis. Energy Sources 28:933–940

Demirbas A (2007) Thermal degradation of fatty acids in biodiesel production by supercritical methanol. Energy Explor Exploit 25:63–70

Demirbas A (2009a) Production of biodiesel from algae oils. Energy Sour Part A Recovery, Utilization Environ Effects 31:163–168

Demirbas A (2009b) Biodiesel from waste cooking oil via base-catalytic and supercritical methanol transesterification. Energy Convers Manag 50:923–927

Dickinson J, Jackson T, Matthews M, Cripps A (2009) The economic and environmental optimisation of integrating ground source energy systems into buildings. Energy 34:2215–2222

Divakaran R, Pillai VNS (2002) Flocculation of algae using chitosan. J Appl Phycol 14:419–422. doi:10.1021/es902405a

Dote Y, Sawayama S, Inoue S, Minowa T, Yokoyama SY (1994) Recovery of liquid fuel from hydrocarbon rich microalgae by thermochemical liquefaction. Fuel 73:1855–1857

Doucha J, Straka F, Líivanský (2005) Utilization of flue gas for cultivation of microalgae (*Chlorella* sp.) in an outdoor open thin-layer photobioreactor. J Appl Phycol 17:403–412

Douskova I, Doucha J, Machat J, Novak P, Umysova D, Vitova M, Zachleder V (2008) Microalgae as a means for converting flue gas CO_2 into biomass with a high content of starch. Bioenergy: challenges and opportunities international conference and exhibition on bioenergy. Guimarães, Portugal, April 6th–9th

Drapcho CM, Nhuan NP, Walker TH (2008) Biofuels engineering process technology, Mc Graw Hill, New York

Edwards M (2010) Algae World 2010. Industry survey. Report in association with the centre of management technology

Edzwald JK (1993) Algae, bubbles, coagulants, and dissolved air flotation. Water Sci Technol 27:67–81

Ehimen EA, Sun ZF, Carrington CG (2010) Variables affecting the in situ transesterification of microalgae lipids. Fuel 89:677–684

Encinar JM, Beltran FJ, Ramiro A, Gonzalez JF (1998) Pyrolysis/gasification of agricultural residues by carbon dioxide in the presence of different additives: influence of variables. Fuel Process Technol 55:219–233

Engler CR (1993) Cell breakage. In: Harrison RG (ed) Protein purification process engineering. CRC Press, London, pp 37–55

Eriksen N (2008) The technology of microalgal culturing. Biotechnol Lett 30:1525–1536

Eriksen N, Riisgard F, Gunther W (2007) On-line estimation of O_2 production, CO_2 uptake, and growth kinetics of microalgal cultures in a gas-tight photobioreactor. J Appl Phycol 19:161–174

FAO (2007) Food and agriculture organization of the United Nations. http://www.fao.org

Flynn KJ (2001) A mechanistic model for describing dynamic multi-nutrient, light, temperature interactions in phytoplankton. J Plankton Res 23:977–997

Flynn KJ (2003) Modelling multi-nutrient interactions in phytoplankton; balancing simplicity and realism. Prog Oceanogr 56:249–279

Flynn KJ (2008a) Use, abuse, misconceptions and insights from quota models—the droop cell quota model 40 years on. Oceanography Mar Biol Annu Rev 46:1–23

Flynn KJ (2008b) The importance of the form of the quota curve and control of non-limiting nutrient transport in phytoplankton models. J Plankton Res 30:423–438

Gallagher BJ (2011) The economics of producing biodiesel from algae. Renew Energy 36:158–162

Gao S, Yang J, Tian J, Ma F, Tu G, Du M (2010a) Electro-coagulation–flotation process for algae removal. J Hazard Mater 177:336–343

Gao S, Du M, Tian J, Yang J, Ma F, Nan J (2010b) Effects of chloride ions on electro-coagulation–flotation process with aluminum electrodes for algae removal. J Hazard Mater 182:827–834

Ghirardi ML, Kosourov S Tsygankov A, Seibert M (2000) Two-phase photobiological algal H_2-production system. In: Proceedings of the 2000 U.S. DOE hydrogen program review. National Renewable Energy Laboratory, Golden, Colorado, pp 1–13

Goldemberg J (2007) Ethanol for a sustainable energy future. Science 315:808–810

Gouveia L, Oliveira AC (2009) Microalgae as a raw material for biofuels production. J Ind Microbiol Biotechnol 36:269–274

Gouveia L, Nobre BP, Marcelo FM, Mrejen S, Cardoso MT, Palavra AF, Mendes RL (2007) Functional food oil coloured by pigments extracted from microalgae with supercritical CO_2. Food Chem 101:717–723

Gouveia L, Marques AE, Lopes da Silva T, Reis A (2009) Neochloris oleabundans UTEX #1185: a suitable renewable lipid source for biofuel production. J Ind Microbiol Biotechnol 36:821–826

Greenwell HC, Laurens LML, Shields RJ, Lovitt RW, Flynn KJ (2010) Placing microalgae on the biofuels priority list: a review of the technological challenges. J R Soc Interface 7:703–726

Gregoire-Padro CE (2005) Hydrogen basics. First annual international hydrogen energy implementation conference. The New Mexico Hydrogen Business Council, Santa Fe, NM

Grierson S, Stezov V, Ellem G, Mcgregor R, Herbertson J (2008) Thermal characterization of microalgae under slow pyrolysis conditions. J Anal Appl Pyrol 85:118–123

Gudin C, Chaumont D (1991) Cell fragility—the key problem of microalgae mass production in closed photobioreactors. Bioresour Technol 38:145–151

Guerrero M (2009) Producción de Aceites a partir de Microalga. Seminario Internacional de Biocombustibles de Algas. Antofagasta, Chile, 7–8 October

Hankamer B, Lehr F, Rupprecht J, Mussgnug JH, Posten C, Kruse O (2007) Photosynthetic biomass and H_2 production by green algae: from bioengineering to bioreactor scale-up. Physiol Plant 131:10–21

Hansel A, Lindblad P (1998) Towards optimization of cyanobacteria as biotechnologically relevant producers of molecular hydrogen, a clean and renewable energy source. Appl Microbiol Biotechnol 50:153–160

Harun R, Danquah MK (2011) Influence of acid pre-treatment on microalgal biomass for ethanol production. Process Biochem 46:304–309

Harun R, Singh M, Forde GM, Danquah MK (2010a) Bioprocess engineering of microalgae to produce a variety of consumer products. Renew Sust Energ Rev 14:1037–1047

Harun R, Danquah MK, Forde GM (2010b) Microbial biomass as a fermentation feedstock for bioethanol production. J Chem Technol Biotechnol 85:199–203

Hatti-Kaul R, Mattiasson B (2003) Release of protein from biological host. In: Hatti-Kaul R, Mattiasson B (eds) Isolation and purification of proteins. CRC, London, pp 1–28

He H, Feng C, Huashou L, Wenzhou X, Yongjun L, Yue J (2010) Effect of iron on growth, biochemical composition and paralytic shellfish poisoning toxins production of *Alexandrium tamarense*. Harmful Algae 9:98–104

Heasman M, Diemar J, O'Connor W, Sushames T, Foulkes L (2000) Development of extended shelf-life microalgae concentrate diets harvested by centrifugation for bivalve molluscs—a summary. Aquacult Res 31:637–659

Hills FJ, Johnson SS, Geng S, Abshahi A, Peterson GR (1983) Comparison of four crops for alcohol yield. Calif Agric 37:17–19

Hirano A, Ryohci U, Shin H, Yasuyuki O (1997) CO_2 fixation and ethanol production with microalgal photosynthesis and intracellular anaerobic fermentation. Energy 22:137–142

Hirano AK, Hon-Nami K, Kunito S, Hada M, Ogushi Y (1998) Temperature effect on continuous gasification of microalgal biomass: theoretical yield of methanol production and its energy balance. Catal Today 45:399–404

Ho S-H, Chen W-M, Chang J-S (2010) *Scenedesmus obliquus* CNW-N as a potential candidate for CO_2 mitigation and biodiesel production. Bioresour Technol 101:8725–8730

Hon-Nami K (2006) A unique feature of hydrogen recovery in endogenous starch-to-alcohol fermentation of the marine microalga. Appl Bioch Biotech 129–132:808–828

Hon-Nami K, Kunito S (1998) Microalgae cultivation in a tubular bioreactor and utilization of their cells. Chin J Oceanol Limnol 16:75–83

Hopkins T (1991) In: Seetharam and Sharma (eds) Purification and analysis of recombinant proteins. Marcel Dekker, New York

Hsue HT, Chu H, Chang CC (2007) Identification and characteristics of a Cyanobacterium Isolated from a Hot spring with dissolved inorganic carbon. Environ Sci Technol 41:1909–1914

Hu DW, Liu H, Yang CL, Hu EZ (2008a) The design and optimization for light-algae bioreactor controller based on artificial neural network-model predictive control. Acta Astronaut 63:1067–1075

Hu Q, Sommerfeld M, Jarvis E, Ghirardi M, Posewitz M, Seibert M, Darzins A (2008b) Microalgal triacylglycerols as feedstocks for biofuel production: perspectives and advances. Plant J 54:621–639

Huntley ME, Redalje DG (2007) CO_2 mitigation and renewable oil from photosynthetic microbes: a new appraisal. Mitig Adapt Strat Glob Change 12:573–608

IEA (2006) World energy outlook 2006. International Energy Agency, Paris

Iliopoulou EF, Antonakou EV, Karakoulia SA, Vasalos IA, Lappas AA, Triantafyllidis KS (2007) Catalytic conversion of biomass pyrolysis products by mesoporous materials: effect of steam stability and acidity of Al-MCM-41 catalysts. Chem Eng J 134:51–57

Illman AM, Scragg AH, Shales SW (2000) Increase in *Chlorella* strains calorific values when grown in low nitrogen medium. Enzyme Microb Technol 27:631–635

IPCC (2007) Intergovernmental panel on climate change 'AR4 synthesis report'. http://www.ipcc.ch. Cited 30 Nov

Iwasaki I, Hu Q, Kurano N, Miyachi S (1998) Effect of extremely high-CO_2 stress on energy distribution between photosystem I and photosystem II in a 'high-CO_2' tolerant green alga, *Chlorococcum littorale* and the intolerant green alga *Stichococcus bacillaris*. J Photochem Photobiol B 44:184–190

Jena U, Das KC (2009) Production of biocrude oil from microalgae via thermochemical liquefaction process. Bioenergy Engineering, Bellevue, Washington, DC, BIO-098024. American Society of Agricultural and Biological Engineers, St. Joseph, Michigan, 11–14 October

John RP, Anisha GS, Nampoothiri KM, Pandey A (2010) Micro and macroalgal biomass: a renewable source for bioethanol. Bioresour Technol. doi:10.1016/j.biortech.2010.06.139

Jones KO, Clarkson N, Young A J (2002) Intelligent process modelling of a continuous algal production system. In: Dochain D, Perrier M (eds) Computer applications in biotechnology.

2001 8th IFAC international conference on computer applications in biotechnology (CAB8), pp 239–243. Pergamon, New York

Kapdan IK, Kargi F (2006) Bio-hydrogen production from waste materials. Enzyme Microb Technol 38:569–582

Khan SA, Rashmi Hussain MZ, Prasad S, Banerjee UC (2009) Prospects of biodiesel production from microalgae in India. Renew Sust Energ Rev 13:2361–2372

Kheshgi HS, Prince RC, Marland G (2000) The potential of biomass fuels in the context of global climate change. Annu Rev Energy Environ 25:199–244

Khotimchenko SV, Yakovleva IM (2005) Lipid composition of the red alga *Tichocarpus crinitus* exposed to different levels of photon irradiance. Phytochemistry 66:73–79

Kim J, Kang C, Park T (2006) Enhanced hydrogen production by controlling light intensity in sulfur-deprived *Chlamydomonas reinhardtii* culture. Int J Hydrogen Energy 31:1585–1590

Knuckey RM, Brown MR, Robert R, Frampton DMF (2006) Production of microalgal concentrates by flocculation and their assessment as aquaculture feeds. Aquacult Eng 35:300–313

Koopman B, Lincoln EP (1983) Autoflotation harvesting of algae from high-rate pond effluents. Agr Wastes 5:231–246

Krohn BJ, McNeff CV, Yan B, Nowlan D (2011) Production of algae-based biodiesel using the continuous catalytic Mcgyan process. Bioresour Technol 102:94–100

Kula MR, Schütte H (1987) Purification of proteins and the disruption of microbial cells. Biotechnol Progr 3:31–42

Kurano N, Ikemoto H, Miyashita H (1995) Fixation and utilization of carbon dioxide by microalgal photosynthesis. Energy Convers Manag 36:689–692

Lardon L, Elias T, Sialve B, Esteyer J-F, Bernard O (2009) Life-cycle assessment of biodiesel production from microalgae. Environ Science Tech 43:6475–6481

Lazarova V, Phillippe R, Sturny V, Arcangell JP (2006) Evaluation of economic viability and benefits of urban water reuse and its contribution to sustainable development. Water Prac Technol 1:1–11

Lee DH (2011) Algal biodiesel economy and competition among bio-fuels. Bioresour Technol 102:43–49

Lee SJ, Kim SB, Kim JE, Kwon GS, Yoon BD, Oh HM (1998) Effects of harvesting method and growth stage on the flocculation of the green alga *Botryococcusbraunii*. Lett Appl Microbiol 27:14–18

Lehmann J (2007) A handful of carbon. Nature 447:143–144

Li Y, Horsman M, Wu N, Lan CQ, Dubois-Calero N (2008) Biofuels from microalgae. Biotechnol Prog 24:815–820

Liang Y, Beardall J, Heraud P (2006) Changes in growth, chlorophyll fluorescence and fatty acid composition with culture age in batch cultures of *Phaeodactylum tricornutum* and *Chaetoceros muelleri* (Bacillariophycee). Bot Mar 49:165–173

Lindblad P, Christensson K, Lindberg P, Fedorov A, Pinto F, Tsygankov A (2002) Photoproduction of H_2 by wildtype *Anabaena* sp. PCC 7120 and a hydrogen uptake deficient mutant: from laboratory experiments to outdoor culture. Int J Hydrogen Energy 27:1271–1281

Lopes da Silva T, Reis A, Medeiros R, Oliveira AC, Gouveia L (2009) Oil production towards biofuel from autotrophic microalgae semicontinuous cultivations monitorized by flow citometry. Appl Biochem Biotechnol 159:568–578

Lopez CVG, Fernandez FGA, Sevilla JMF, Fernandez JFS, Garcia MCC, Molina Grima E (2009) Utilization of the cyanobacteria *Anabaena* sp ATCC 33047 in CO_2 removal processes. Biores Technol 100:5904–5910

Lueschen W, Putnam D, Kanne B, Hoverstad TA (1991) Agronomic practices for production of ethanol from sweet sorghum. J Prod Agric 4:619–625

Lv P, Yuan Z, Wu C, Ma L, Chen Y, Tsubaki N (2007) Biosyngas production from biomass catalytic gasification. Energy Convers Manage 48:1132–1139

Lynch DV, Thompson GA (1982) Low temperature-induced alterations in the chloroplast and microsomal membranes of *Dunaliella salina*. Plant Physiol 69:1369–1375

Madamwar D, Garg N, Shah V (2000) Cyanobacteria hydrogen production. World J Microb Biotechnol 16:757–767

Maeda K, Owada M, Kimura N, Omata K, Karube I (1995) CO_2 fixation from the flue gas on coal-fired thermal power plant by microalgae. Energy Convers Manag 36:717–720

Marris E (2006) Black is the new green. Nature 442:624–626

Masukara H, Nakamura K, Mochimaru M, Sakurai H (2001) Photohydrogen production and nitrogenase activity in some heterocystous cyanobacteria. Biohydrogen II:63–66

Mata TM, Martins AA, Caetano NS (2010) Microalgae for biodiesel production and other applications: a review. Renew Sust Energ Rev 14:217–232

Matsumoto M, Hiroko Y, Nobukazu S, Hiroshi O, Tadashi M (2003) Saccharification of marine microalgae using marine bacteria for ethanol production. Appl Bioch Biotcch 105:247–254

Maxwell EL, Folger AG, Hogg SE (1985) Resource evaluation and site selection for microalgae production systems. SERI/TR-215-2484

Meier D, Faix O (1999) State of the art of applied fast pyrolysis of lignocellulosic materials—a review. Bioresour Technol 68:71–77

Melis A, Happe T (2001) Hydrogen production. Green algae as a source of energy. Plant Physiol 127:740–748

Melis A, Zhang L, Forestier M, Ghirardi ML, Seibert M (2000) Sustained photobiological hydrogen gas production upon reversible inactivation of oxygen evolution in the green alga *Chlamydomonas reinhardtii*. Plant Physiol 122:127–136

Mendes RL (2008) Supercritical fluid extraction of active compounds from algae. In: Supercritical fluid extraction of nutraceuticals and bioactive compounds, pp 189–213

Meng X, Yang J, Xu X, Zhang L, Nie Q, Xian M (2009) Biodiesel production from oleaginous microorganisms. Rev Energy 34:1–5

Miao X, Wu Q (2004) High yield bio-oil production from fast pyrolysis by metabolic controlling of *Chlorella protothecoides*. J Biotechnol 110:85–93

Miao X, Wu Q (2006) Biodiesel production from heterotrophic microalgal oil. Bioresour Technol 97:841–846

Miao X, Wu Q, Yang C (2004) Fast pyrolysis of microalgae to produce renewable fuels. J Anal Appl Pyrolysis 71:855–863

Middelberg APJ (1995) Process-scale disruption of microorganisms. Biotechnol Adv 13:491–555

Miller SA (2010) Minimizing land use and nitrogen intensity of bioenergy. Environ Sci Technol 44:3932–3939

Mohan D, Pittman CU, Steele PH (2006) Pyrolysis of wood/biomass for bio-oil: a critical review. Energy Fuel 20:848–889

Moheimani NR, Borowitzka MA (2006) The long-term culture of the coccolithophore *Pleurochrysiscarterae* (Haptophyta) in outdoor raceway ponds. J Appl Phycol 18:703–712

Molina Grima E, Fernandez F, Camacho F (1999) Photobioreactors: light regime, mass transfer and scale up. J Biotechnol 70:231–247

Molina Grima ME, Belarbi EH, Fernandez FGA, Medina AR, Chisti Y (2003) Recovery of microalgal biomass and metabolites: process options and economics. Biotechnol Adv 20:491–515

Molina Grima E, Acién Fernández FG, Robles Medina A (2004) Downstream processing of cell-mass and products. In: Richmond A (ed) Handbook of microalgal culture: biotechnology and applied phycology. Blackwell, London, pp 215–252

Mollah MYA, Morkovsky P, Gomes JAG, Kesmez M, Parga J, Cocke DL (2004) Fundamentals, present and future perspectives of electrocoagulation. J Hazard Mater 114:199–210

Morowvat MH, Rasoul-Amini S, Ghasemi Y (2010) *Chlamydomonas* as a "new" organism for biodiesel production. Bioresour Technol 101:2059–2062

Morweiser M, Kruse O, Hankamer B, Posten C (2010) Developments and perspectives of photobioreactors for biofuel production. Appl Microbiol Biotechnol 87:1291–1301

Mulbry W, Kondrad S, Buyer J (2008a) Treatment of dairy and swine manure effluents using freshwater algae: fatty acid content and composition of algal biomass at different manure loading rates. J Appl Phycol 20:1079–1085

Mulbry W, Kondrad S, Pizzarro C, Kebede-Westhead E (2008b) Treatment of dairy effluent using freshwater algae: algal productivity and recovery of manure nutrients using algal turf scrubbers. Bioresour Technol 99:8137–8142

Murakami M, Ikenouchi M (1997) The biological CO_2 fixation and utilization project by RITE (2): screening and breeding of microalgae with high capability in fixing CO_2. Energy Convers Manag 38(suppl 1):S493–S497

Mussatto SI, Dragone G, Guimarães P, Silva JP, Carneiro LM, Roberto IC, Vicente A, Domingues L, Teixeira JA (2010) Technological trends, global market, and challenges of bio-ethanol production. Biotechnol Adv 28:817–830

Mussgnug JH, Klassen V, Schlüter A, Kruse O (2010) Microalgae as substrates for fermentative biogas production in a combined biorefinery concept. J Biotechnol 150:51–56

Nobre B, Marcelo F, Passos R, Palavra A, Gouveia L, Mendes R (2006) Supercritical carbon dioxide extraction of astaxanthin and other carotenoids from the microalga *Haematococcus pluvialis*. Eur Food Res Technol 223:787–790

Oh HM, Lee SJ, Park MH, Kim HS, Kim HC, Yoon JH, Kwon GS, Yoon BD (2001) Harvesting of *Chlorella vulgaris* using a bioflocculant from *Paenibacillus* sp. AM49. Biotechnol Lett 23:1229–1234

Ohlrogge J, Allen D, Berguson B, DellaPenna D, Shachar-Hill Y, Stymne S (2009) Driving on biomass. Science 324:1019–1020

Oilgae (2009) Report. http://www.oilgae.com. Accessed Nov 2010

Oilworld (2009) http://www.oilworld.biz. Accessed Nov 2010

Okabe K, Murata K, Nakanishi M, Ogi T, Nurunnabi M, Liu Y (2009) Fischer–Tropsch synthesis over Ru catalysts by using syngas derived from woody biomass. Catal Lett 128:171–176

Ono E, Cuello JL (2006) Feasibility assessment of microalgal carbon dioxide sequestration technology with photobioreactor and solar colector. Biosyst Eng 95:597–606

OriginOil (2010) http://www.originoil.com November 2010

Ozgener O, Hepbasil A (2007) A review on the energy and exergy analysis of solar assisted heat pump systems. Renew Sust Energ Rev 11:482–496

Pakdel H, Roy C (1991) Hydrocarbon content of liquid products and tar from pyrolysis and gasification of wood. Energy Fuels 5:427–436

Park JBK, Craggs RJ, Shilton AN (2011) Wastewater treatment high rate algal ponds for biofuel production. Bioresour Technol 102:35–42

Patil V, Tran KQ, Giselrod HR (2008) Towards sustainable production of biofuel from microalgae. Int J Mol Sci 9:1188–1195

Perner-Nochta I, Posten C (2007) Simulations of light intensity variations in photobioreactors. J Biotechnol 131:276–285

PetroAlgae (2010) http://www.petroalgae.com. Accessed Nov 2010

Petrusevski B, Bolier G, van Breemen AN, Alaerts GJ (1995) Tangential flow filtration: a method to concentrate freshwater algae. Water Res 29:1419–1424

Pienkos PT, Darzins A (2009) The promise and challenges of micro-algal derived biofuels. Biofuel Bioproducts Biorefin 3:431–440

Poelman E, DePauw N, Jeurissen B (1997) Potential of electrolytic flocculation for recovery of micro-algae. Resour Conserv Recyc 19:1–10

Powell EE, Hill GA (2009) Economic assessment of an integrated bioethanol–biodiesel–microbial fuel cell facility utilizing yeast and photosynthetic algae. Chem Eng Res Des 87:1340–1348

Price GD, Woodger FJ, Badger MR, Howitt SM, Tucker L (2004) Identification of a SulP-type bicarbonate transporter in marine cyanobacteria. Proc Natl Acad Sci U S A 101:18228–18233

Prins MJ, Ptasinski KJ, Janssen FJJG (2006) More efficient biomass gasification via torrefaction. Energy 31:3458–3470

Proviron (2010) http://www.proviron.com/algae. Accessed Nov 2010

Pulz O (2001) Photobioreactors: production systems for phototrophic microorganisms. Appl Microbiol Biotehnol 57:287–293

Pulz O (2004) Valuable products from biotechnology of microalgae. Appl Microbiol Biotechnol 65:635–648

Pushparaj B, Pelosi E, Torzillo G, Materassi R (1993) Microbial biomass recovery using a synthetic cationic polymer. Elsevier, Oxford, Royaume Uni

Qiang H (2004) Environmental effects on cell composition. In: Richmond A (ed) Handbook of microalgal culture: biotechnology and applied phycology. Wiley-Blackwell, New York, pp 83–93

Raison JK (1986) Alterations in the physical properties and thermal responses of membrane lipids: correlations with acclimation to chilling and high temperature. In: St Joh JB, Berlin E, Jackson PG (eds) Frontiers of membrane research in agriculture. Rowman and Allanheld, Totowa, pp 383–401

Ran CQ, Chen ZA, Zhang W, Yu XJ, Jin MF (2006) Characterization of photobiological hydrogen production by several marine green algae. Wuhan Ligong Daxue Xuebao 28(suppl 2): 258–263

Ranga Rao A, Dayananda C, Sarada R (2007a) Effect of salinity on growth of green alga *Botryococcus braunii* and its constituents. Bioresour Technol 98:560–564

Ranga Rao A, Sarada R, Ravishankar G (2007b) Influence of CO_2 on growth and hydrocarbon production in *Botryococcus braunii*. J Microbiol Biotechnol 17:414–419

Rao KK, Hall DO (1996) Hydrogen production by cyanobacteria: potential, problems and prospects. J Mar Biotechnol 4:10–15

Reijnders L (2009) Microalgal and terrestrial transport biofuels to display fossil fuels. Energies 2:48–56

Reijnders L, Huijbregts MAJ (2009) In: Transport biofuels: a seed to wheel perpertive. Springer, London

Richmond A (2004) Handbook of microalgal culture: biotechnology and applied phycology. Blackwell Science, New York

Rodolfi L, Zitelli GC, Bassi N, Padovani G, Biondi N, Bonini G, Tredici MR (2009) Microalgae for oil: strain selection, induction of lipid synthesis and outdoor mass cultivation in a low-cost photobioreactor. Biotech Bioeng 102:100–112

Rosenberg JN, Oyler GA, Wilkinson L, Betenbaugh MJ (2008) A green light for engineered algae: redirecting metabolism to fuel a biotechnology revolution. Biotechnology 19:430–436

Rossi N, Jaouen O, Legentilhomme P, Petit I (2004) Harvesting of cyanobacterium *Arthospira platensis* using organic filtration membranes. Food Bioprod Process 82:244–250

Rossignol N, Vandanjon L, Jaouen O, Quemeneur F (1999) Membrane technology for the continuous separation microalgae/culture medium: compared performances of cross flow microfiltration and ultrafiltration. Aquacult Eng 20:191–208

Rubio J, Souza ML, Smith RW (2002) Overview of flotation as a wastewater treatment technique. Miner Eng 15:139–155

Sanchez F, Vasudevan PT (2006) Biodiesel production by enzymatic transesterification of olive oil. Appl Biochem Biotechnol 135:1–14

Schenk PM, Skye R, Thomas-Hall, Stephens E, Marx UC, Mussgnug JH, Posten C, Kruse O, Hankamer B (2008) Second generation biofuels: high-efficiency microalgae for biodiesel production. Bioenergy Res 1:20–43

Schütz K, Happe T, Troshina O, Lindblad P, Leitão E, Oliveira P, Tamagnini P (2004) Cyanobacterial H_2-production—a comparative analysis. Planta 218:350–359

Shen Y, Pei Z, Yuan W, Mao E (2009) Effect of nitrogen and extraction method on lipid yield. Int J Agric Biol Eng 2:51–57

Shiraiwa Y, Goyal A, Tolbert NE (1993) Alkalization of the medium by unicellular green algae during uptake of dissolved inorganic carbon. Plant Cell Physiol 34:649–657

Sialve B, Bernet N, Bernard O (2009) Anaerobic digestion of microalgae as a necessary step to make microalgal biodiesel sustainable. Biotechnol Adv 27:409–416

Singh A, Nigam PS, Murphy JD (2011) Renewable fuels from algae: an answer to debatable land based fuels. Bioresour Technol 102:10–16

Solix_Biofuels (2010) http://www.solixbiofuels.com. Accessed Nov 2010

Stehfest K, Toepel J, Wilhelm C (2005) The application of micro-FTIR spectroscopy to analyze nutrient stress-related changes in biomass composition of phytoplankton algae. Plant Physiol Biochem 43:717–726

Stephenson AL, Kazamia E, Dennis JS, Howe CJ, Scott SA, Smith AG (2010) Life-cycle assessment of potential algal biodiesel production in the United Kingdom: a comparison of raceways and air-lift tubular bioreactors. Energy Fuels 24:4062–4077

Stern N (2006) The economics of climate change. HM Treasury, London

Subhadra B, Edwards M (2010) An integrated renewable energy park approach for algal biofuel production in United States. Energy Policy 38:4897–4902

Subitec (2010) http://www.subitec.com. Accessed Nov 2010

Sukenik A, Shelef G (1984) Algal autoflocculation—verification and proposed mechanism. Biotechnol Bioeng 26:142–147

Sukenik A, Bilanovic D, Shelef G (1988) Flocculation of microalgae in Brackish and sea waters. Biomass 15:187–199

Sukenik A, Yamaguchi Y, Livne A (1993) Alterations in lipid molecular species of the marine eustigmatophyte *Nannochloropsis* sp. J Phycol 29:620–626

Sveshnikov D, Sveshnikova N, Rao K, Hall D (1997) Hydrogen metabolism of mutant forms of *Anabaenavariabilis* in continuous cultures and under nutritional stress. FEMS Microbiol Lett 147:297–301

Tabak J (2009) Biofuels. Infobase Publishing, New York

Tamagnini P, Leitão E, Oliveira P, Ferreira D, Pinto F, Harris D, Heidorn T (2007) Cyanobacterial hydrogenases. Diversity, Regulation and Applications. FEMS Microbiol Rev 31:692–720

Tickell J (2000) From the fryer to the fuel tank. The complete guide to using vegetable oil as an alternative fuel, Tallahasseee, USA

Uduman N, Qi Y, Danquah MK, Forde GM, Hoadley A (2010) Dewatering of microalgal cultures: a major bottleneck to algae-based fuels. J Renew Sustain Energy 2:012701

Ueda R, Hirayama S, Sugata K, Nakayama H (1996) Process for the production of ethanol from microalgae. US Patent 5,578,472

Ueno Y, Kurano N, Miyachi S (1998) Ethanol production by dark fermentation in the marine green alga, *chlorococcum littorale*. J Ferment Bioeng 86:38–43

Um BH, Kim YS (2009) Review: a chance for Korea to advance algal-biodiesel technology. J Ind Eng Chem 15:1–7

USDA (2007) US Department of Agriculture. http://www.usda.gov

van Beilen JB (2010) Why microalgal biofuels won't save the internal combustion machine. Biofuels Bioprod Biorefin 4:41–52

van Harmelen T, Oonk H (2006) Microalgae biofixation processes: applications and potential contributions to greenhouse gas mitigation options. TNO Built Environmental Geosciences, Apeldoorn

Vandamme D, Foubert I, Meesschaert B, Muylaert K (2010) Flocculation of microalgae using cationic starch. J Appl Phycol 22:525–530

Vergara-Fernandez A, Vargas G, Alarcon N, Velasco A (2008) Evaluation of marine algae as a source of biogas in a two-stage anaerobic reactor system. Biomass Bioenergy 32:338–344

Wang Z, Pan Y, Dong T, Zhu X, Kan T, Yuan L, Torimoto Y, Sadakata M, Li Q (2007) Production of hydrogen from catalytic steam reforming of biooil using C12A7-O-based catalysts. Appl Catal A 320:24–34

Wang B, Li Wu N, Lan CQ (2008) CO_2 bio-mitigation using microalgae. Appl Microbiol Biotechnol 79:707–718

Weissman JC, Tillett DM (1992) Aquatic species project report; NREL/MP-232-4174. In: Brown LM, Sprague S (eds) National Renewable Energy Laboratory, pp 41–58

Weissman J, Goebel RP, Benemann JR (1988) Photobioreactor design: mixing, carbon utilization and oxygen accumulation. Biotechnol Bioeng 31:336–344

Westfalia (2010) http://www.westfalia-separator.com/en/about/aboutnews/newspressedetail.php? ID=1021. Accessed Nov 2010

Wilde EW, Benemann JR, Weissman JC, Tillett DM (1991) Cultivation of algae and nutrient removal in a waste heat utilization process. J Appl Phycol 3:159–167

Williams JA (2002) Keys to bioreactor selection. Chem Eng Prog 98:34–41

Wu Q, Miao X (2003) A renewable energy from pyrolysis of marine and freshwater algae. Recent Adv Mar Biotechnol Biomater Bioprocess 111–125

Xu H, Miao X, Wu Q (2006) High quality biodiesel production from a microalga *Chlorella prototheicoides* by heterotrophic growth in fermenters. J Biotechnol 126:499–507

Yang Z, Guo R, Xu X, Fan Xa, Li X (2010) Enhanced hydrogen production from lipid-extracted microalgal biomass residues through pretreatment. Int J Hydrogen Energy 35:9618–9623

Yang J, Xu M, Zhang X, Hu Q, Sommerfeld M, Chen Y (2011) Life-cycle analysis on biodiesel production from microalgae: water footprint and nutrients balance. Bioresour Technol 102:159–165

Yoon RH, Luttrell GH (1989) The effect of bubble size on fine particle flotation. Miner Process Extract Metal Rev 5:101–122

Zeiler KG, Heacox DA, Toon ST et al (1995) The use of microalgae for assimilation and utilization of carbon dioxide from fossil fuel-fired power plant flue gas. Energy Convers Manag 36:707–712

Zhila NO, Kalacheva GS, Volova TG (2005) Effect of nitrogen limitation on the growth and lipid composition of the green alga *Botryococcus braunii* Kutz IPPAS H-252. Russ J Plant Physiol 52:311–319

Zwart RWR, Boerrigter H, van der Drift A (2006) The impact of biomass pre-treatment on the feasibility of overseas biomass conversion to Fischer–Tropsch products. Energy Fuels 20:2192–2197

Printed by Printforce, the Netherlands